T0186501

EUROPEAN MEETINGS ON EDUCATIONAL RESEARCH

Part A:
Reports of Educational Research Symposia, Colloquies, and Workshops
organised under the auspices of the Council of Europe

Part B:
European Conferences of Directors of Educational Research Institutions
organised in cooperation with UNESCO, the UNESCO Institute
for Education in Hamburg, and the Council of Europe

PART A: VOLUME 26

SOCIALISATION OF SCHOOL CHILDREN
AND THEIR EDUCATION FOR DEMOCRATIC VALUES
AND HUMAN RIGHTS

COUNCIL OF EUROPE – STRASBOURG

Socialisation of School Children and their Education for Democratic Values and Human Rights

REPORT OF THE COLLOQUY OF DIRECTORS OF
EDUCATIONAL RESEARCH INSTITUTIONS HELD IN
ERICEIRA (PORTUGAL) ON 17-20 OCTOBER 1989

Edited by
Hugh Starkey
Westminster College, Oxford

SWETS & ZEITLINGER B.V. AMSTERDAM / LISSE

SWETS & ZEITLINGER INC. BERWYN, PA

PUBLISHING SERVICE

Library of Congress Cataloging-in-Publication Data

Colloquy of Directors of Educational Research Institutions (1989 : Ericeira, Portugal)
　　Socialisation of School Children and their Education for Democratic Values and Human Rights ;
Report of the Colloquy of Directors of Educational Research Institutions held in Ericeira (Portugal)
on 17 - 20 October 1989 / edited by Hugh Starkey.
　　　　p.　　cm. -- European meetings on educational research, ISSN 0924-0578 ; v. 26. Part A,
Reports of Educational Research Symposia, Colloquies, and Workshops organised under the auspices
of the Council of Europe
　　At head of title: Council of Europe --Strasbourg.
　　Includes bibliographical references
　　ISBN 90 265 1148 5
　　　　1. School children--Europe--Social conditions--Congresses. 2. Socialization--Congresses.
3. Education--Europe--Aims and objectives--Congresses. 4. Human rights--Europe--Congresses.
5. Educational innovations--Europe--Congresses. I. Starkey, Hugh. II. Council of Europe.
III. Title. IV. Title: Socialisation of school children and their education for democratic values and
human rights. V. Series: European meetings on educational research ; v. 26 VI. Series: European
meetings on educational research. Part A, Reports of Educational Research Symposia, Colloquies,
and Workshops organised under the auspices of the Council of Europe

　　Lc206.E8C65 1989
　　370.19--dc20　　　　　　　　　　　　　　　　　　　　　　　　　　90-24323
　　　　　　　　　　　　　　　　　　　　　　　　　　　　　　　　　　　CIP

CIP-gegevens Koninklijke Bibliotheek, Den Haag

Socialisation

Socialisation of school children and their education for democratic values and human rights : report of
the colloquy of directors of educational research institutions held in Ericeira (Portugal) on 17-20
october1989 / ed. Hugh Starkey. – Amsterdam [etc]: Swets & Zeitlinger – (European meetings in
educational research. ISSN 0924-0578 : vol. 26 pt. A)
Met lit. opg.
ISBN 90-265-1148-5
SISO 450.49　UDC [37.018.2:341.231.14]+316.6–053.5 NUGI 724
Trefw.: mensenrechten in het onderwijs / socialisatie ; schoolkinderen.

Cover design: Rob Molthoff
Cover printed in the Netherlands by Casparie IJsselstein
Printed in the Netherlands by Offsetdrukkerij Kanters B.V., Alblasserdam

ISBN 90 265 1148 5
ISSN　0924-0578
NUGI 724

Contents

Preface

The Colloquy held in Ericeira (Portugal) was one of a series of educational research meetings which have become an important element in the programme of the Council for Cultural Co-operation of the Council of Europe since 1975. European co-operation in educational research aims at providing Ministries of Education with research findings so as to enable them to prepare their policy decisions. Co-operation should also lead to a joint European evaluation of certain educational developments.

The Colloquies bring together directors of educational research institutions and other research workers from a selected number of the 27 countries taken part in the work of the Council for Cultural Co-operation. The purpose is to compare research findings and experience on a particular topic of current interest; to identify areas of research so far neglected and to discuss possibilities for joint research projects. The reports, as well as a selection of the papers of these meetings, are usually published a a book so that Ministries and research workers, as well as a wider public (parents, teachers, press), are kept informed of the present state of research at European level.

The theme "research into the socialisation of school children and their education for democratic values and human rights in education at school" was chosen because the Council for Cultural Co-operation thought it would be appropriate to devote a Colloquy to this subject in 1989 on the occasion of the 200th Anniversary of the Declaration of the Rights of Man and the Citizen.

The Colloquy was organised by the Portuguese Ministry of Education (Institute for Educational Innovation) in co-operation with the Council for Cultural Co-operation of the Council of Europe. Six papers (covering France, Federal Republic of Germany, Portugal, Sweden, Switzerland) were presented in plenary session and then discussed in three working groups. National and individual reports from various countries, as well as lists of research projects and bibliographies, were tabled as background material. On the final day, the Rapporteur General, Mr. Hugh STARKEY, summed up the situation and the conclusions emerging from the Colloquy.

The following countries were represented: Belgium, Cyprus, Denmark, Finland, France, Federal Republic of Germany, Greece, Holy See, Iceland, Italy, Luxembourg, Malta, Norway, Portugal, San Marino, Sweden, Switzerland, Turkey, United Kingdom, Yugoslavia. There were also observers from Guinea-Bissau, Sao-Tomé e Principe, USA and several teacher organisations (WCOPT, FIPESO, FIAI, FISE, SPIE). The list of participants is given at the end of this book.

The Council of Europe is particularly grateful to Professor Dr. Manuel FERREIRA PATRICIO for having chaired the meeting and to the Institute for Educational Innovation (Dr. Teresa Quintela, Maria Emilia Apolinario) for their excellent work in preparing and organising the Colloquy. The Council of Europe would also like to thank the lecturers, the group chairmen, the rapporteurs and Mr. Hugh STARKEY for having written the final report and done the editing.

Strasbourg, 1 July 1990
Michael VORBECK
Head of the Section for Educational Research & EUDISED

8

Part 1
Report and
Commissioned Papers

1.1
Report

Hugh Starkey
Rapporteur Général
United Kingdom

1.1.1 SOME MAJOR POINTS FROM THE COMMISSIONED PAPERS PRESENTED: INDICATIONS OF DIFFERENT LINES OF RESEARCH

1.1.1.1 Professor Bartolo Paiva Campos, Portugal

Young people in Portugal under 30, surveys show, have a relatively low level of political participation (one in eight had attended a union meeting) although nearly half belong to an association of some sort (e.g. a sports club). However, this is not necessarily significantly worse than the record of the population in general.

The values of young people tend to be more personal and associated with immediate gratification in the less well educated. Students and the better educated are likely to have wider social values. However, there is a trend whereby personal values appear increasingly dominant over social ones.

Education can thus be shown to help social awareness, but even well-educated young people have been shown to lack co-operative strategies in interpersonal conflicts.

Schools' attempts to educate for personal and social values and for democracy are hampered, may studies show, by the hidden curriculum. Knowledge is transmitted, but without the experience of responsibility that makes it real. The management of the school is exclusively in the hands of the teachers, denying students precious opportunities to develop skills of organisation and management.

Fear of bias and indoctrination caused the demise of the first generation of political education in Portugal (1974-76). The new reforms have re-introduced personal, social and civic education (see the report of the IEI below).

Considerable research has been undertaken into strategies for socio-moral education building on Kohlberg. A number of experiments in active learning and through special projects have been undertaken.

In conclusion, although schools cannot be expected to take on the entire process of democratic socialisation, young people can be led to greater social consciousness through a successful school experience. This is likely to be enhanced if the school community develops structures and practices supportive of co-operative learning. For instance:
– moving from academic study to challenging and meaningful project-based work
– giving time to reflect as well as to act
– developing social and psychological skills rather than merely transmitting models of orthodox behaviour and values
– creating conditions for developing good personal relationships.

Contact: Dr. Bartolo PAIVA CAMPOS, Faculdade de Psicologia et de Ciencias de Educação, Universidade do Porto, Rua das Taipas 76, P-4000 PORTO.

1.1.1.2 Dr. Bengt Thelin, Sweden

Swedish schools are expected to be supportive of democracy and human rights. This is to be seen in an international or global perspective, not merely a European one. The Brundtland Report to the United Nations (Our Common Future, 1988) outlines dangers to the planet and the need for international measures. One of these measures is education for global survival.

Dr Thelin outlined the world situation as *unique* in that there is now the capacity to destroy the planet. In many respects the situation is *absurd,* in that the comparison between resources spent on armaments and those spent on health and education cannot be justified by any logic. The question of global responsibility is particularly *relevant* to the generation in school, as they will have longer to live with the consequences.

Education for global responsibility involves action skills as well as knowledge. Swedish schools have been involved with local community twinnings with schools or towns in the third world. It is common for NGOs to work with schools, for instance through school clubs, but also working directly with teachers on some courses. This action approach is important to avoid feelings of helplessness in young people faced with huge global issues.

Young people learn about global issues via the media, particularly television. They receive very fragmented messages. The role of education is to give frameworks which can make sense of these fragments. During the discussion period it was reported that the experience of the University of Padua supports this analysis. A global approach to human rights issues is well received by young people.

Contact: Dr. Bengt THELIN, Swedish National Board of Education, Karlavägen 108, S- 106 42 STOCKHOLM.

1.1.1.3 Dr. Heinz Schirp, Federal Republic of Germany

The research project is based on Kohlberg's work on stages of moral development. A grid, giving the six stages of moral judgment, is the basic theoretical framework. The first stage is taking place in three secondary schools (a Hauptschule, a Realschule and a Gymnasium). Teachers are introduced to the concepts of moral-cognitive development and a teaching strategy is adopted based on moral dilemmas. Through discussing these dilemmas students are led to understand that moral responses can be made at different levels. They can then be encouraged to aspire to a higher level.

A number of curriculum areas lend themselves to this approach, namely:
– German (literature)
– History (development of values-systems)
– Social and political education (values and vested interests)
– Biology (environment, health, people and nature)
– Geography (responsibility for the environment)

Trial schools arrange school community meetings where participating students can discuss problems and conflicts arising from school life. They are encouraged and helped to take responsibility for the meetings and the outcomes.

The whole project is being carefully observed and monitored. Many lessons are being recorded or video-ed. Although it is too early for definitive conclusions, a number of useful preliminary observations can be made:
– Teachers have responded well to the concept of moral-cognitive development. It provides them with a pedagogical instrument and helps

them to understand their students better.

- Both social and academic skills are necessary and links must be made between the two.
- Democratic education can be compatible with subject studies.
- Unless the school adopts a Community Education approach, the constraints of school organisation may be too restrictive for the project to succeed fully.

A human rights perspective corresponds in this analysis with a high level (stage 6) of moral judgment. However, it is possible to lead students towards this level by starting with cases of which they have first hand experience. Thus a discussion of women's rights can start with sex specific socialisation in family and school, lead on to discrimination at work and move to a discussion of woman's roles and rights in other cultures. This final stage involves generalising to the point where universal principles are considered.

Dr Schirp stressed the need to work from the level of the student, which implies making the discussions relevant. The school must itself be prepared to promote democracy, so that the content of lessons is coherent with the ethos of the school. Teachers need to be introduced to the theoretical framework used and be encourage to use participatory teaching styles.

Contact: Dr Heinz SCHIRP, Landesinstitut für Schule and Weiterbildung, Paradieser Weg 64, D-4770 SOEST.

1.1.1.4 Mr Jacques-André Tschoumy, Switzerland

It is possible to identify a number of obstacles to the effective implementation of human rights education. These obstacles can be classified as educational, ideological, philosophical and social.

Educational obstacles include developing a democratic teaching style and making this acceptable to teachers and parents. It involves a great effort of initial and inservice teacher education and is thus to be seen in the long term. Teaching for human rights is also interdisciplinary, whereas schools are predominantly subject dominated.

The ideological obstacle is linked with teachers perceptions of their role. For many there is a need or obligation to be neutral; to transmit knowledge but not attitudes. There clearly is a link between morality, human rights and politics. Human rights education is, however, more than just moral education.

The philosophical obstacle is to do with the notions of universality and indivisibility in human rights. These concepts are at odds with schools and

14

society which compartmentalise life rather than allow for a holistic approach.

The social obstacle is to do with the fact that human rights are of society and not of the state. Many areas of society have not developed along democratic lines as political systems have. The economy, for instance, is based on a utilitarian rather than liberal logic. Analyses referring to socio-economic classes are to be seen alongside socio-spatial groupings.

In the discussion that followed the paper, Professor Tarrow from California reminded the meeting of the paradox of human rights education having a high level of apparent legitimacy but a relatively low level of implementation. American research into educational innovation shows that new programmes are best implemented when there is support from the top but also from those involved on the ground. Recent work on human rights education in Latin America showed disappointing results because the teachers were unhappy with the materials. It is now important to develop formative and summative tools to measures the effectiveness of human rights education programmes.

Contact: M. Jacques-André TSCHOUMY, Institut Romand de Recherche et de Documentation Pédagogiques, Faubourg de l'Hôpital, 43 CH-2007 Neuchatel 7

1.1.1.5 M. François Audigier, France

The most important and largest action-research programme on human rights education in Europe took place in France from 1984-87 and it is now in its dissemination phase. The project involved some 100 teachers in seven large teaching institutions. It was coordinated by the INRP, the national educational research institute.

Within each establishment an interdisciplinary team was created. They worked to establish a body of knowledge that could be taught within the constraints of the French education system. The research has been published and additional material, such as resource guides and annotated bibliographies is also available.

Amongst the major conclusions, the project noted the following:
1. The necessity for a solid and clearly presented body of knowledge based on official texts.
2. That any attempt at human rights education runs into obstacles and difficulties inherent in the controversial nature of the subject.
3. Since human rights is both a legal and an ethical concept this has implications for the school as a society. The knowledge transmitted may conflict with certain practices and procedures at school.

Recent research into history teaching has revealed that students have great difficulty in developing a complex view of history. Simplistic, binary thinking may often lead to quite erroneous conclusion such as being unable to understand that prior to the Revolution the French bourgeoisie were rich, but without political power. Teachers and students avoid treating the period of the Terror, perhaps because to understand it would be considered as starting to justify it.

Human rights education involves developing new school structures. Students may be involved in drawing up their own codes of behaviour, although it is then necessary to consider the question of power and what happens if such a code is breached. Teachers need help in enabling conflicts to be resolved non-violently. In fact teacher training and retraining is an essential element in a successful human rights education programme. This training should also include an element of education in the law.

Contact: M. François AUDIGIER, INRP, 29 rue d'Ulm, F-75230 PARIS Cedex 05.

The report of the project is published by INRP at the above address. It is called "Education aux droits de l'homme" in the Collection "Rapports de recherches", 1987 no. 13.

1.1.2 MAIN CONCLUSIONS OF THE WORKING GROUPS

1.1.2.1 Some unresolved questions noted by the groups

i. The extent to which the school is able to interact with other agencies (family, media, NGOs) that are also responsible for human rights education. Defining the specific contribution of the school.

ii. The need to agree and clarify the terminology of human rights education. Definition is still required for: democracy; the democratic school; participation; socialisation; third generation rights. The expression "human rights" has both a legal and a moral sense and researchers need to avoid ambiguity.

iii. Even where an apparently democratic and participative model for schools exists, the reality is often disappointing.

iv. There are areas where the cultural or political values of certain groups appear to conflict with human rights values. It is important to identify successful strategies for defusing such conflicts.

v. The complexity of the area and the need to develop research instruments both suggest that it is unrealistic to expect immediate results from this research.

vi. Human rights education through participation often entails the setting up of special projects involving a temporary flexibility of the timetable.

1.1.2.2 Conclusions reached based on research evidence

Successful programmes are likely to have the following characteristics:
– support from the top and from the base
– content that is both local and global
– effective and continuing inservice education for teachers
– teams of teachers in individual schools
– supporting networks between schools
– involvement of the community, parents and administrators
– schools involved in a programme of self-evaluation
– effective dissemination strategies.

1.1.2.3 Research needs and directions

i. The cataloguing of assessment and evaluation instruments and performance indicators. The development of new and validated instruments suitable for human rights education.

ii. The creation of data-bases logging current human rights education programmes in various parts of Europe and their accompanying materials and resources.

iii. Studies of good practice in human rights education, including curriculum materials and methods but also effective structures for participation by students and models for inservice education. Evaluations should include student responses.

iv. Creating action-research models using material deriving from the studies of good practice.

v. A survey of which states and authorities have official guidelines or policies including references to human rights and democratic values.

vi. Research comparing the behaviour and attitudes of young people in countries with formal civic education programmes with that of young people where there is no such provision.

vii. Studies of what factors make for the effective delivery of interdisciplinary programmes, such as human rights education.

viii. Studies of experience in initial teacher education together with a programme to develop and evaluate a human rights element in teacher training.

ix. Studies and evaluations of project based approaches to human rights education.

x. Case studies of schools and authorities which have taken combating racism as the focal point of their human rights education.

1.1.3 TEN KEY IDEAS DEVELOPED AT THE COLLOQUY

1.1.3.1 Global perspective

Participants insisted on the necessity for a planetary perspective to human rights education. They also noted the need for a global or whole-school approach.

1.1.3.2 Dissemination

Much work has already been done and is on-going. Results and good practice need to be made known to the widest possible public. As well as the education press and journals, video and television programmes should be used. The Committee of Ministers' Recommendation R(85)7 on teaching and learning about human rights in schools should be disseminated to initial teacher training institutions and to educational research workers as well as to teachers.

1.1.3.3 Motivation

Human rights education has been found to motivate students and teachers to work hard and give of their best.

1.1.3.4 Climate

The climate of the school has been shown to affect students' learning. This effect can be negative of positive. There is an immediate need to develop performance indicators for school climate.

1.1.3.5 Stages

It is tempting, but simplistic to identify ages at which particular approaches are appropriate. Although primary education is likely to focus on co-operation and tolerance, children can, of course, also be developing action skills and acquiring knowledge and concepts. The interesting work using stages of moral-cognitive development is a worthwhile line of research.

1.1.3.6 Law

Few teachers have detailed knowledge of the law, yet this is a key concept in human rights education. The development and evaluation of programmes of law in education is to be encouraged.

1.1.3.7 Cross-curricular themes

Building interdisciplinary teams is necessary for effective human rights education, at least at secondary level, as the French project demonstrates. The French national curriculum guidelines provide one outstanding model for an interdisciplinary approach. Other models should be developed to suit other national criteria.

1.1.3.8 Participation

This was agreed to be an essential component and aim, but research is still needed on how to measure it and how to evaluate its effectiveness. Experience of different models should be made available and evaluated.

1.1.3.9 Better schools

Participants were convinced that schools adopting a human rights ethos are likely to have a peaceful and stimulating climate and well motivated pupils. It remains to be demonstrated that such schools are more effective in an instrumental sense.

1.1.3.10 Europe

Europe in its widest sense is developing a new unity based on shared democratic values. Education has a crucial role to play in this. One positive step would be the setting up of a European education in human rights research network. This should be augmented by national or regional centres for human rights education.

1.1.4 FURTHER ACTION

Participants agreed to form an informal *network for human rights education* informing each other of interesting pilot experiments, research projects and publications. Dr. H. SCHIRP (Federal Republic of Germany) tabled a questionnaire for the purpose of collecting such data. It was suggested that all should make an effort to compile relevant data and send them to Dr. Schirp for preliminary analysis. At a later stage, this material would be reproduced and disseminated by the Council of Europe.

Mr Vorbeck drew participants' attention to the computerised EUDISED database containing descriptions of educational research and development projects; EUDISED might help to make relevant projects more widely known.

1.1.5 APPENDIX: RESEARCH REPORTED TO THE COLLOQUY

1.1.5.1 Belgium

Whilst not specifically concerned with human rights education, a large-scale and long-term project (1973-) on the renewal of the primary school (RPS) provides valuable evidence about how schools successfully achieve innovation.

The RPS project has involved hundreds of schools in the Flemish speaking part of Belgium. It was initiated by a high powered national team including universities and the ministry. Schools implementing the programme were expected to implement a local strategy and these were categorised and evaluated as follows:

Innovation through PLANNING: A plan including specific indications of desired changes in teaching practice is formulated and management effort is applied to its implementation. Many positive changes are likely to occur in a short time.

Innovation through INTERACTION. Frequent discussions and consultations within the school and with external facilitators encourage and support the involvement of all team members in the innovation. Existing structures, such as weekly meetings, are exploited. Written information is exchanged. Many changes can be observed in a short time.

Innovation emphasising RISK AVOIDANCE. This slow, steady, cautious approach attempts to achieve consensus on the changes and to attempt only those elements that are seen to be realizable in the short term. The effect of a heavy emphasis on discussion and information may lead to very little change in the early stages at least.

Innovation through CO-OPTION. In this model there is heavy reliance on an external change agent. Unless the school develops a collective response, staff members may not take responsibility for their own development and innovations are likely to be short lived.

Further research identified key features of successful innovative schools, namely:
– A school leadership with clear goals and able to articulate a vision.
– Working as a team towards the goals.
– Professional interest amongst the staff and good communications and support.
– Trust and a supportive network.

Contact: Prof. Roland VANDENBERGHE, Centre for Educational Policy and Innovation, Katholieke Universiteit, Vesaliusstraat, B-3000 LEUVEN.

1.1.5.2 Cyprus

Educational research in Cyprus is the responsibility of the Group for Educational Research at the Pedagogical Institute. The main priority is research into the four main curriculum subjects, namely: Greek, Maths, Science and English. The Institute also assists schools and teachers engaged in action research projects.

A major and influential study of the values and attitudes of young people was published in 1982.

Research specifically on human rights education has concentrated on the results of the introduction of a new civics course in 1983.

Teachers have been evaluating the new learning styles promoted by the course including discussion activities, project work, field work, questionnaires and other means of data collection, self evaluation and the study of values and priorities in the thoughts and actions of peoples and individuals.

Contact: Dr. Panayoitis PERSIANIS, Director, Pedagogical Institute, NICOSIA.

1.1.5.3 Finland

Research into "the formation of a geographical world view" published by the Finnish UNESCO Commission, notes that seventh grade students and even fifth graders express ethical attitudes in their study of geography. They express a strong sense of justice and disapproval of violations of human rights, even when the cases are far removed in space of time. The report concludes that such case-studies may give young people the opportunity to develop ethical positions, whereas controversial cases nearer home might raise problems with parents and teachers.

Further research on moral development suggests that boys are able to generalise better than girls, who tend to limit their moral horizon to their immediate surroundings. Bullies and teasers were found to have either no moral perspective or one no higher than the conventional morality. Another substantial study mapped young people's world view and their view of themselves and of human beings.

Human rights and education in Finland

The aims of secondary education as revised in 1986 include:
- young people as individuals and members of society should be capable of cooperation and have a will for peace
- education should encourage both a sense of national culture and a readiness to undertake international cooperation.

The Finnish UNESCO Commission's seminar in November 1988 concluded:
- because human rights is not a separate subject, but a theme within other subjects, it may not be present within many school districts
- even where human rights are incorporated into the aims of the school, the teaching and climate of the school may work against the aims being realised
- discrimination exists and teachers' knowledge of human rights is inadequate.

A pilot scheme launched in 1986 with high level support involved 11 schools and 2000 students in trialling materials and methods for human rights education. These are currently being produced in English.

In 1988 the National Board of Education published a guide to international education using human rights ethics as its basis. In September 1989 a European seminar on Education for Human Rights was held in Finland under the auspices of the national UNESCO Commission. A report will be published.

A survey on teacher education revealed that in 1986/87 only 10% of teachers in training (and only 2% of secondary trainees) studied international education and only half the modules offered in these programmes dealt with human rights. As a result a new research and development project has been initiated in the Autumn of 1989, at the University of Jyväskylä, on "the inclusion of international education in the training of subject teachers".

Contact: Dr Annikki JÄRVINEN, Institute for Educational Research, Seminaarinkatu 15, SF-40100 JYVÄSKYLÄ.

1.1.5.4 Greece

Since the 1985 reform, human rights is formally included in the curriculum of the upper secondary school. Specific reference is made to international declarations and conventions in history, sociology and civic education.

In June 1988 a principal working group on the whole curriculum was formed, and this has been examining human rights education. It has been influenced by the work of UNESCO and a Unesco sponsored seminar was held in March 1989. This concluded that:
- human rights education should be a theme in all subject areas not a separate subject.
- the teaching should have a practical emphasis but avoid dogmatism
- further research and development would be undertaken at three levels, namely: school life; subject studies; teacher training.

The seminar provided many suggestions for the integration of human rights within subject studies in the Greek curriculum. The climate appears to be favourable for these to be put into practice.

Human rights is included in university teacher training courses by reference to the philosophy of Rawls and the developmental work of Kohlberg and Piaget.

Contact: Prof. Myrto DRAGONA-MONAHOU, (University of Crete) 18, P Tsaldari Maroussi, GR-15122 ATHENS.

1.1.5.5 Holy See

"Youth and Peace"
Research conducted among a sample of 13,053 young people aged 19 in 12 European Countries (1985-89) by the Institute of Education at the Pontifical Salesian University.

The survey found that nearly half of the respondents, and over two thirds in Catholic schools, considered that school had stimulated them to take an interest in peace. This is at least as important an influence as the family or associations. Usually the interest was stimulated by discussions, debates or meetings organised by the school. Less than a third of students considered that their school encouraged participation in outside activities related to peace. When asked to relate each of list of eleven words to the concepts of violence, "school" was perceived as being relatively far from violence. On the other hand "education" and "democracy" were closely related to the word peace.

Three quarters of those questioned thought education for peace to be "a valid strategy for the maintaining of peace among the different social groups of a nation". Southern European respondents considered this highly feasible, but English students were very dubious.

The same proportion also felt that education itself is a means to promoting peaceful relationships between individuals.

Contact: Rev Father Prof. Guilielmo MALIZIA, Università Pontificia Salesiana, Piazza del Ateneo Salesiano 1, I-00139 Roma.

1.1.5.6 Italy

The Ministry of Education is committed to a continuous effort to introduce a human rights element into civic education courses at primary and secondary level.

At the same time non-governmental organisations including the churches have recently set up courses and training centres for the socio-political education of young people. Courses cover human rights, peace, democracy, the environment and European integration.

Local authorities have also taken initiatives in this area, notably Padua which for the past two years has run a course for 300 young people entitled "The Experience of Democracy".

Following the 1974 UNESCO recommendation, various regional authorities have adopted policies to promote culture and peace, founded on the respect for human rights.

Since 1988 the University of Padua's Study and Training Centre on Human and Peoples' Rights has been running a three year graduate in-service course to train teachers and school planning experts in human rights. Arising from this is a new research project on 'Peace and human rights culture: teaching and research' which has been adopted by the National Research Council. The Centre has a national research network involving 20 universities. Its current programme includes:
– the international dimension of the content of human rights education
– the interdisciplinary dimension of human research and education.

Educational researchers tend to concentrate on a specific aspect of human rights education rather than human rights as a guiding principle. For instance research is being undertaken into:
– updating formal civics courses
– political education
– intercultural education in a multicultural society
– development education
– environmental education.

Much of this research stems from personal commitment on the part of researchers and their involvement with NGOs. There is thus interaction between universities, schools and society.

Contact: Prof. Antonio PAPISCA, Centre for the Study of Human and Peoples' Rights, Università di Padova, Via del Santo 28, I-3513 PADOVA.

1.1.5.7 The Netherlands
Abstracts of several projects were provided. One study had reached conclusions as follows.

The reproduction of racism in social studies textbooks:
Ethnic minority groups were mentioned in about half of the books studied. The books tended to give a one-sided, unfavourable and stereotyped image of minority groups. It is concluded that such books contribute to the reproduction of racism in society.

Another project provides a set of questions it is seeking to answer.

Prejudices against ethnic minorities:
The project seeks to arrive at a clear definition of prejudice, discover how prejudices develop and map the extent of racial prejudice amongst different age groups and in different schools. It also examines the role of the school and how it can help to combat prejudice, including a study of indicator of prejudice and their validity.

Contact: Institute for Educational Research in the Netherlands (SVO), Sweelinckplein 14, NL-2517 GK 'S-GRAVENHAGE.

1.1.5.8 Northern Ireland (United Kingdom)

"Value education in a divided society"
A number of projects are being undertaken at the Centre for the Study of Conflict at the University of Ulster. All are investigating the new phenomenon of the integrated school in an area where the education system is generally divided on sectarian lines. Studies are investigating school links, the philosophy and curriculum of integrated schools, the roles of parents and teachers in such schools and the attitudes of potential parents.

Contact: Seamus DUNN and Alan SMITH, Centre for the Study of Conflict, University of Ulster, COLRAINE, BT52 1SA, Northern Ireland.

1.1.5.9 Portugal

The Ericeira Colloquy was hosted by the recently formed Institute for Educational Innovation (IEI). This was formed in 1986 coinciding with a major educational reform. This reform is itself influenced by and is the expression of human rights principles such as:
– education for autonomy and freedom
– education for democracy
– education for development
– education for solidarity
– education for change

Amongst the projects of the Institute is the promotion and evaluation of a new kind of school under the Cultural School Project. The project promotes extra-curricular activities which take place on school premises and which complement and interact with the formal curriculum. 68 schools are currently involved with the project. They are given resources, including a time-allowance to develop extra-curricular activities such as: civil/cultural enrichment; sports and physical education; artistic education; links with the community.

The schools involved have shown increased motivation in all groups of students, even those with little academic success. A recent census identified 656 newly started clubs including: consumer's rights, community action; ecology; intercultural activities; European club; journalism; school radio.

The IEI also monitors innovations and projects around the country. It identified a large number of human rights related events in primary and secondary schools in 1989. These included exhibitions commemorating the 40th anniversary of the Universal Declaration and the bicentenary of the French Revolution; activities linked with UNICEF, UNESCO and Amnesty International; a seminar on children's rights; human rights as a topic in school magazines and newspapers.

A major project involving the IEI is monitoring and helping schools integrate children of recently returned emigrants.

The IEI is not alone in promoting human rights education. In June 1989 the Government introduced moral and civic education in all years of compulsory schooling. This is, however, set against Catholic religious education and students must choose one or the other.

The new education reform puts particular emphasis on the integration of children with special needs. All children are seen to have a right to an education adapted to their specific needs.

In late 1988 the Portuguese Ministers of Justice and Education set up a Commission for the Protection of Human Rights and Equality in Education. The Commission acknowledges the fundamental importance of human rights in education and stresses the identical rights of all citizens. It recommends compulsory civic education. The Commission promotes human rights and civic education in schools and is working to eliminate discrimination on the grounds of sex, class, economic status or ethnic origin.

Contact: Prof. Dr. Manuel FERREIRA PATRICIO, Director, Instituto de Inovação Educacional, Travessa das Terras de Sant'Ana, 15, P-1200 Lisboa.

1.1.5.10 San Marino

The reform of secondary education in San Marino has involved schools opening out to society and to the world of work. In conjunction with Italian universities training programmes have been developed to prepare Heads for a more active, less bureaucratic role. Care has been taken to provide appropriate and adequate resources for the reform.

San Marino offered to host a future follow-up colloquy to that held in Ericeira.

Contact: Mme Carla NICOLINI, Preside, Scuola Secondaria Superiore Statale, Contrada Santa Croce, RSM-47031 CITTÀ DI SAN MARINO.

1.1.5.11 Sweden

An action programme for the Internationalisation of education was published by the National Board of Education (NBE) in 1989:

"Random surveys undertaken by the NBE .support the assumption that instruction concerning human rights is both a difficult and neglected field. Clearer guidance needs to be given on this point in the curricula. UNESCO and the Council of Europe have taken active steps to strengthen the position of teaching about human rights. There is scope here for better compliance on Sweden's part. Above all, international humanitarian law appears to be unknown."

"In the commentaries now being prepared on the teaching of general subjects in compulsory school, special emphasis will be put on the internationalisation aspect."

"Where upper secondary school is concerned, a new draft civics syllabus has been presented to the Government. International questions have been allotted more prominent role in this draft document..."

"Furthermore, civics will be made a compulsory subject for the experimental three year vocational lines of upper secondary school."

"International education and comparative educational research are amply represented in Sweden. Educational development and, above all, educational research have shown less interest in the great issues concerning the future and the survival of mankind and the position and treatment of those questions in schools."

"... over the next few years the NBE for its part will be trying to devise procedures for encouraging and supporting research and development in education relating to what are usually termed issues of destiny and survival."

Contact: Dr Bengt THELIN, Swedish National Board of Education, Karlavägen 108, S-106 42 STOCKHOLM

1.1.5.12 Turkey

Education is free and compulsory in Turkey and it is seen as the basis for democracy and good citizenship. Equality of education is guaranteed, irrespective of religion or race. In 1985 a thorough curriculum review was undertaken by committees of experts. This process is still underway.

The Turkish education system is an important instrument of socialisation. One question currently under discussion in the Ministry of Education is how to reconcile the expression of a range of values within schools with this normative socialising function.

Contact: Prof. Esin KAHIA, Aü Dil, Tarih, Congrapfya Fakultesi, University of Ankara, TR-ANKARA.

1.1.5.13 United Kingdom and Ireland

"The European Studies (Ireland and Great Britain Project)"
The project started in 1986 and is now in two parts, one for students aged 11-16 another for 16-18 year olds. The 11-16 project involves 2000 students in 18 schools in three countries, Ireland, Northern Ireland and England sharing information for their study of history, geography and personal and social education.

The programme of study is a joint one over three years taught as 7 modules (3 history, 3 geography and one PSE). Information is exchanged by surface mail, by E-mail and by residential courses.

The project stresses a particular learning style, characterised as follows:
– a climate of caring and mutual respect
– a sense of achievement for each individual
– cooperative learning rather than a competitive atmosphere
– increased group responsibility for decisions taken
– deliberation and reflection to help students identify their own learning.

The team reports a 98% satisfaction rate among the participating students. The project work approach and the links with other schools have proved very motivating.

The project has been extended to involve a older age-group and further countries, initially Belgium and then France and West Germany.

Contact: Dr Roger AUSTIN, European Studies Project, Ulster Folk and Transport Museum, 153 Bangor Road, Cultra, BT18 OEU, Northern Ireland.

1.1.5.14 Yugoslavia

A report was received from the Institute for Pedagogical Research at the University of Zagreb.

Research is being undertaken in a number of relevant areas such as:
– child-centred learning and the acquisition of critical thinking
– resocialisation of pupils with special learning needs
– the social and educational status and rights of minorities, migrants and
  marginalised populations.
Findings are expected in 1990.

Contact: Dr Zlata GODLER, Faculty of Philosophy, University of Zagreb, Hrelinska 21, YU-41000 ZAGREB.

1.2
Reports of the Working Groups

1.2.1 ENGLISH SPEAKING GROUP

Chairperson: Dr. Claire BURSTALL
Rapporteur: Professor Kenneth WAIN

1.2.1.1 Issues raised by the group

Apart from the matters listed under *1.2.1.2 General conclusions and recommendations* and *1.2.1.3 Specific recommendations*, the following are the main matters raised in the course of the group meetings:

- To what extent, it was asked, should human rights education be the responsibility and function of the school, and how much should it be a matter of the home?

- The question was raised about women and minority rights – to what extent *do* they currently feature in human rights curricula? To what extent *should* they feature?

- What is the precise relationship between democracy and human rights?

Is a democratic education the same as an education for human rights? How do you evaluate its effects?

- What should be the status of second and third generation rights in the curriculum? Should we start with the global perspective recommended by Dr. Thelin?

- The question about socialisation in human rights raised the problem of defining "socialisation". There are probably other problems of definition in human rights discourse among researchers.

- To what extent do democratic societies need democratic schools? What do we mean by democratic schools? What are democratic competences and how are they relevant to human rights education?

- The problem of clarifying our verbal usages applies most basically to the term human rights itself; researchers should be clear whether they mean legal rights or moral when they use it. Though both dimensions are important, human rights education should begin with the moral aspect.

- If the agreed rights are those set out in the European Convention, to what extent are they embodied in schools and their curricula? How does the democratic environment of a school affect human rights education?

- The best time to implement human rights teaching appears to be in the school at the earliest age.

1.2.1.2 General recommendations

- There is need for comparative research projects through sociological and psychological approaches, on a European scale, into the social demands for human rights education. The object would be to determine the attitudes of different parties (teachers, parents, pupils) towards human rights and towards being educated in a human rights morality, and to seek explanation for these attitudes.

- There is need for research into situations where conventional cultural values conflict with human rights, the object being to study how the conflict can be overcome.

- It is important to bring home to the policy-makers the difficulties inherent in evaluating human rights education programmes; the considerable demand for human and financial resources, the sensitivity and complexity of the area, and the lack of appropriate methodologies for carrying out such research. In the light of these facts expecting immediate results is unrealistic.

- In connection with the previous point, it is important to highlight the dearth of appropriately validated evaluation instruments for human rights education programmes. The task of creating such instruments should be regarded as a high priority and should be tackled at the national and European levels.

– Another project should be compiling studies of good schooling practices for human rights education, where good means: (a) passing on the appropriate knowledge; (b) developing the right attitudes and fostering the right values; (c) facilitating action.

– Participation is crucial to the human rights programmes, but the notion raises different research issues that have not been addressed properly so far, about the meaning of participation, the degree that should be allowed at different levels of schooling, its precise educative values, the skills it requires if it is to be successful and how these can be taught.

– In-service teacher training (INSET) is important for the implementation of human rights education, for both teachers and heads or principals. There is a need to identify models of good practice where they exist.

– There is need for a programme to work on an all-European curriculum strategy for the teaching of human rights which would serve as a reference point for local curricula and which ensure the broader European dimension on the issues, desired by the Committee of Ministers (see Recommendation No. R (85) 7).

– There is need also for research into new ways of getting research findings on human rights education to teachers, utilising audiovisual aids and readily accessible reading material alongside the conventional reports and journals.

– The role of the head of school in the context of a school structure and environment geared towards democratic and human rights education, needs to be reassessed in the light of his or her traditional role, but this raises the question of the politic of the school into which research needs to be made for models and answers.

1.2.1.3 Specific recommendations

– We need to know how human rights education programmes are being assessed where they *are* being implemented, and what the obstacles are where they where they are *not* being implemented.

– A remark was made about "collective pessimism" among young people about the world and their future in it. Is it true? How widespread is it? How can it be combated through human rights teaching? There should be the appropriate research to find answers to these questions.

- Related to this point is the question of the self-perception of young people. Do they see themselves as having the means and power to influence events on a personal and a wider plan? It not, why not? Again there is need for research in this area, which should be school based.

- There is a need to create action-research models utilising the material gathered in the fifth general recommendation for teachers involved in human rights education and for curriculum development in the area.

- A crucial area of immediate research which should not be difficult but which is of the first importance, should be on teacher training institutions – into what they are currently doing, if anything, to prepare teachers to be favourably disposed towards human rights education and equipped to handle it.

- A database on currently functioning human rights education projects with full information is a must. This can be created through a questionnaire survey.

- A research project should be set up without delay to begin the necessary process of instrument development referred to in the fourth general recommendation.

1.2.2 FRENCH SPEAKING GROUP

Chairman and Rapporteur: Dr. Georges WIRTGEN

1.2.2.1 Introduction

Education for democratic values is important for the socialisation of children, in particular in Western societies with multi-ethnic communities.

Education in this sense should, however, not be left to the school alone, but member government of the Council of Europe should involve a wide range of social and political groups in an overall strategy.

1.2.2.2 How to implement education for democratic values

Legal framework

To be effective education for democratic values should be based on legal regulations (acts of parliament, regulations, e.g. those of Geneva).

Research suggested.
- Survey to find out which states have adopted laws defining the objectives of teaching about democratic values.

– Evaluation of the educational results obtained in these countries compared to others.

Continuity of education

Education of this kind should involve all levels of education, from pre-school to university. Nevertheless the main focus of information should be the 12-15 age group. At this age youngsters become aware of social values and are, from a cognitive point of view, able to accept and understand them and to compare them with their own life experience.

In higher education more in-depth human rights education may be necessary for certain professions (e.g. doctors, psychologists, teachers).

Research suggested
– Survey to find out whether adolescents in countries where there is time given to structured human rights education have different attitudes or behaviour in civic and social matters, as opposed to young people in countries with no formal provision.

Interdisciplinary themes

The situation differs from country to country. In some countries education for democratic values is a subject in its own right, in others it forms part of a wider subject such as moral and social education, in others it is taught across several subjects.

An interdisciplinary approach seems preferable to that of a separate teaching unit.

– Following the French example, one may find objectives related to democratic values in a number of subjects.

– On the other hand, certain problems areas might be dealt with in the form of overall themes cutting across several disciplines. In these cases the teachers concerned will have to co-ordinate their work. Throughout the school year certain periods of time might be set aside for educational projects of this type.

Research suggested
– Evaluation of the French curriculum and similar curricula in other countries.

– Studies of the factors that make successful interdisciplinary curricula. These should be undertaken by research workers and in co-operative with practitioners (teachers and pupils).

Learning through life experience

Several research projects in the field of moral development have illustrated that the correlation between moral knowledge, the level of moral judgment and relevant behaviour is rather weak. It is, therefore, not enough just to teach democratic values in order to make children and adolescent accept and practise them in everyday life. They must be given the opportunity to practise them at school, too.

Certainly there are possibilities for pupils to take part in the school's decision-making, but it is often difficult to make use of them because tradition constitutes a major obstacle.

a. A *project-based approach* helps children to choose the content and questions to be studied.

b. The participation of pupils (or their representatives) in school management is desirable but there is a risk that only certain pupils qualified to become leaders gain advantage from it. Furthermore, the whole model risks becoming a charade of democratic participation if pupils do not hold real power.

c. It is perhaps more realistic to set up a variety of participatory management structures at school (e.g. working parties, clubs) allowing a maximum number of pupils to participate in the school's social life and decision-making.

   The conditions for effective pupil participation are:
   – more flexible timetables;
   – adequate facilities (e.g. rooms).

Research suggested
– Evaluation of ongoing experiments in several countries (e.g. Ireland).

– Survey about schools having experimented with particular forms of pupil participation.

– Study on the conditions required for active pupil participation (flexible timetable, adequate rooms).

– Comparative studies of different models of varying degrees of pupil participation, including also a cost-benefit analysis for education (i.e. practising democratic values).

Freedom of choice

The schools' freedom of choice with regard to projects concerning human rights education, education for democratic values and socialisation is vital for success.

1.2.2.3 Accompanying measures

Teacher education

In item 5 of the Appendix to the Recommendation R (1985) 7 of the Council of Europe's Committee of Ministers (Suggestions for teaching and learning human rights at school), the importance of teacher education has been stressed.

Nevertheless, more effective strategies should be conceived even within traditional academic education, as well as the development of curricula adapted to different specialised training courses. Preparation for the teaching of democratic values has to be envisaged via options adapted to the teacher's professional choice.

Furthermore, both universities and colleges of education should introduce legal, historical and philosophical elements allowing to understand better and to explain afterwards the dynamics of social life and the political basis of international organisations and international law. In order to update and reform in-service education and training of teachers (INSET) it is equally important to *revive the network of information* (e.g. databases) on fundamental data about human rights problems. It is obvious that teachers' education must take account of research and evaluation results.

Research suggested
– Research describing teacher education practice and models in European countries.

– Research on teacher education curricula (development and evaluation).

In order to foster the exchange of information on problems and results in teacher education, The Association for Teacher Education in Europe (ATEE) should be encouraged to set up a European group of teacher trainers asked to examine ways and means for improved teacher education in human rights.

Learning and teaching tools

A European database of teaching material should be set up providing Ministers, schools and teachers with up-to-date information about available learning and teaching aids.

Following commonly agreed standards, the database would serve:
a. teachers;
b. education officials with support from documentation centres which would develop further materials (e.g. folders).

This database would be useful both for teachers and teacher trainees and inform them on the most relevant ongoing activities and educational research projects. This kind of material would be at the disposal of teachers, schools and documentation centres.

These data will constitute a precious source of information on the historical development of educational concepts.

Research

Considering that the human rights issue is a fairly recent one and that concepts and their implementation develop rapidly, educational research in Europe in this field is still in its infancy but needs to be encouraged. Research of this kind will be a key element in INSET.

Based in different European priority centres, research might focus on:
a. Curriculum contents, concepts, legal aspects and participation.
b. Methods suitable for pupils at different ages.
c. Teachers' and pupils' ideas about:
   – new curricula
   – educational strategies developed.
d. Their attitudes and behaviour.
e. Evolution of attitudes and behaviour as a consequence of different strategies.
f. Possibilities and difficulties for schools trying to develop into genuine democratic communities.
g. Attitudes of pupils with regard to one another.

Research would have to be both fundamental research creating new knowledge and action-research helping to improve curricula and to understand the possibilities and limits of this kind of teaching.

1.2.3 MIXED GROUP

Chairperson: Dr. Sigridur VALGEIRSDOTTIR
Rapporteur: Dr. Zlata GODLER

The work of the group centred on four themes:

a. Definition of terms,
b. Sharing of research results,
c. Identification of needed research, and
d. General recommendations.

1.2.3.1 Definition of terms

The need for clarification of terminology and sensitivity to cultural and linguistic interpretation was noted.

1.2.3.2 Research results

Various delegates noted successful human rights and global education programmes characterised by:
– Top-down/Bottom-up strategies of implementation.
– Universal/specific dimensions of content.
– Effective and ongoing teacher in-service education.
– Establishment of educational teams at individual school level.
– Continuing inter-school support networks.
– Involvement of parents, community, teachers, administrators, etc.
– Self-evaluation at individual school level.

1.2.3.3 Research needs

1. The critical need for development and/or identification of appropriate assessment and performance indicators was stressed.

2. In order to establish a database, the group recommends gathering information on:
   – The nature and extent of human rights education where implemented.
   – Available materials and successful teaching methods.
   – Student responses.

3. The group recommends action research on programmes that have taken racism as a focal theme.

4. Case studies of schools with successful democratic structures.

5. Developing performance indicators for school appraisal, including, for instance:
   – number participating in civic education programmes;
   – extent of contacts with community groups;
   – international contacts.

1.2.3.4 General recommendations

1. Sending out the Recommendation R (1985) 7 of the Committee of Ministers of the Council of Europe to teacher training institutions and governments.

2. Ongoing series of seminars on HRE for teacher trainers.

3. Forming of groups composed of concerned, interested people willing to initiate and promote HRE.

1.3

The Institute for Educational Innovation and Human Rights Education in Portugal

Manuel Ferreira Patricio
Institute for Educational Innovation
Portugal

1.3.1 INTRODUCTION

The present document aims at outlining the institutional value of the Institute for Educational Innovation (IEI) and the type of activity it deals with.

Given the theme of this colloquy, it was thought to be a good opportunity to draw attention to initiatives from the IEI connected with human rights and democratic values in the light of the educational reform at the present time.

1.3.2 TERMS OF REFERENCE AND ACTIVITIES OF IEI IN GENERAL

IEI is connected directly to the Ministry of Education, it is defined legally as a body of co-ordination, research and development.

The legislation on the institute (Act No. 3/87 of 3 January 1987) states that the following areas are to be covered:
– the study and development of techniques and pedagogical methods;
– the elaboration and trial of techniques, teaching material and equipment

– mainly for young handicapped integration;
– guidelines and support for establishment providing special education.

IEI was created as a result of Act 46/86 passed on 14 October. This law refers expressly to educational research and innovation. It is easier to understand the function of IEI when taking this law into consideration.

As far as educational research goes, its main objective is to produce scientifically-based evaluations of all ideas put forward on education. This research relies on a support network – higher education institutes with specific centres of departments, and also the setting up of autonomous centres specialised in this filed (Article 50).

As far as innovation goes, the law clearly defines that the government will provide adequate structures, providing support for activities of a curricular nature, increase innovation and appraisal of the educational activity programme. These structures should be worked out in conjunction with schools, educational research institutes and teacher training (Article 52).

The law also refers to research and innovation connected with special education for the underprivileged. The education system should satisfy both the general and specific needs of the students.

IEI has only recently been formed (2 1/2 years ago) and is still in its initial stages. There is a space-factor problem, building repairs are needed. More collaborators are required. The 88/89 budget fell short of needs. However, clear, strategic decisions were made and a fair amount of work was done.

A decisive choice was the bringing of innovation and research much closer together. Organisation, running and operation must have a scientifically sound base. This implies that action in scientific educational research is present as back-up for adjustments, renewals, reform or innovation.

Such a running is essential for the feasibility, practicality and success of IEI.

Another important decision was to connect, as far as possible, the activities of the institute with the present educational reform. As the reform aims to totally restructure the education system at every level – a close relationship must be maintained with the IEI to justify its existence. The need to support ideas with solid scientific fact would have no national institute to turn to, were IEI to be omitted.

On the other hand, once the reform is in force it must be given continuity. Constant reviews will make it up-to-date, otherwise it may become a stagnant fact. Educational reform must be permanently focused on, the objective of the institute would be to keep the educational system in movement, relating research to the educational reality, constantly producing new educational ideas.

The activities of the institute can be divided into three main areas:
– curricular development
– teacher training
– promotion and support of educational innovations in the teaching community of scientifically-based research in a bid to its integration.

The diagrams reproduced below give a succinct idea of our plans.

1.3.3 SECTORS

1.3.3.1 Activities under development in IEI

The following areas are covered by IEI as a central service of research co-ordination and development:
– curricular development
– research/innovation
– teacher training

Each area in turn comprises varying activities, the following are worth mentioning:

– curricular development:
  • starting up of activities, supervision and progressive expansion of the school paradigm behind "Cultural School Project". Multi-faceted school providing well-organised basic structure supported in three ways: (a) the curriculum; (b) extra-curricular activities; (c) interactive element bringing harmony to the two;
  • activities of research and elaboration of educational programmes – (a) measures to improve the quality of both written and spoken Portuguese; (b) measures to improve mathematical concepts and operations (c) the areas connected with expression; (d) elaboration of a pre-school plan and basic education for children with difficulties (both slow learners and mentally retarded ones) – with special emphasis to the initial phase of primary education;

– research/innovation
  • the success of the nationwide competitions on innovative and research projects in education
  • the annual publication of a register of innovative/research projects in education;
  • the journal "Innovation" – a three monthly publication which focuses on incentive education promotes understanding between the scientific and pedagogic communities;

- the elaboration of a catalogue publishing dissertations collected from 1983 onwards (presently being researched);
- the project to build up a central bank of information registering everything pertinent;
- study and research promoted by the IEI through direct contact with researchers – titles such as: "A collection of pedagogy from the Portuguese press; "A critical dictionary of Portuguese education" – or researched by our own institute as "A Portuguese bibliography on education";
- the project of integration Portuguese emigrants' children in the Portuguese educational system (in the process of locating/identifying cases and reasons for failure);

- teacher training
  - the activities related to teacher training, concentrating on continuous training/recycling;
  - analysis related to the "Pre-study of continuous training for pre-school, primary and secondary education" and emphasis on training sessions for teachers involved in "Cultural school project". Based on an experimental programme at the University of Evora, pre-school and primary school teachers are being trained.

The above mentioned projects have the support of the following sectors:
- "Centre of Documentation and Information" plans to develop a coherent and current network of information, regularly updated, with a view to helping current projects;

- "Pedagogical Workshop" provides the technical means in audiovisual/technical/graphic fields, so necessary to the projects;

- "Computer Centre" gives access to the Earn-Bitnet network, connecting the Portuguese scientific/educational community in contact with the rest of the world, providing necessary information.

These activities have taken place based on the approved plan of action. In the absence of an organic diploma, our activities outline the measures felt necessary to be taken by the government, which, understandably varies its direction periodically. The three main areas covered by the institute in August 1988 have been dealt with: curricular development; teacher training; research/innovation. The ministerial project for the organic diploma in June 1989 considers the following: pedagogical innovation; scientifically based educational research; evaluation.

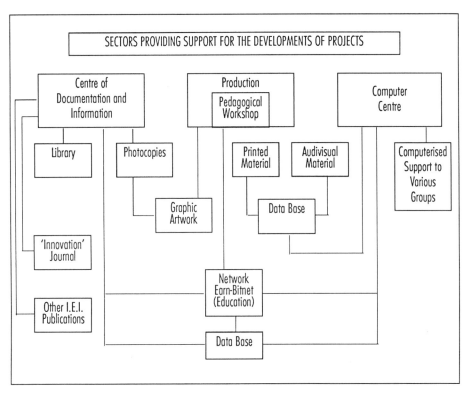

SECTORS PROVIDING SUPPORT FOR THE DEVELOPMENTS OF PROJECTS

Centre of Documentation and Information

Production — Pedagogical Workshop

Computer Centre

Library

Photocopies

Printed Material

Audivisual Material

Computerised Support to Various Groups

Graphic Artwork

Data Base

'Innovation' Journal

Other I.E.I. Publications

Network Earn-Bitnet (Education)

Data Base

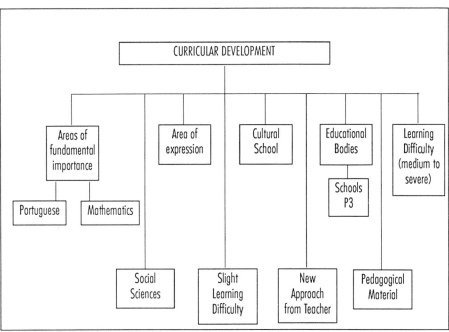

CURRICULAR DEVELOPMENT

Areas of fundamental importance

Area of expression

Cultural School

Educational Bodies

Learning Difficulty (medium to severe)

Portuguese

Mathematics

Schools P3

Social Sciences

Slight Learning Difficulty

New Approach from Teacher

Pedagogical Material

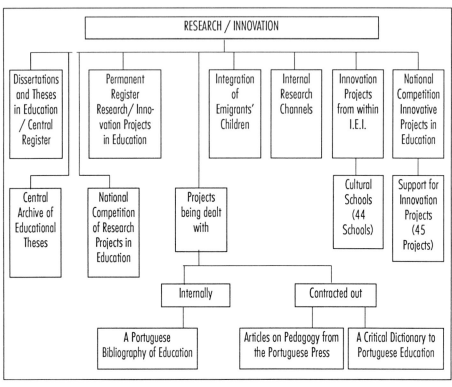

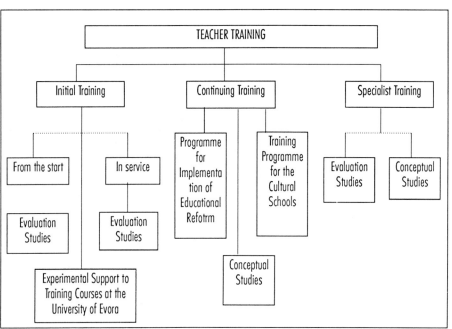

Educational science was only studied at university level in Portugal after 25 April 1974, and it seems to be a widely accepted ideal that Portugal has its own national institute, responsible for promotion and co-ordination of this research also the monitoring of its development. In my point of view this is a fundamental acquisition. It would really help if the above-mentioned areas quoted in the legal documents were taken as mere examples, the totality of the educational reform should be researched, not just particular sectors.

The double function of IEI has now been established: scientifically based educational research providing innovative pedagogical support.

1.3.4 ACTIVITIES OF IEI RELATING TO HUMAN RIGHTS IN EDUCATION

1.3.4.1 The "Education Bill"

The spirit of the Universal Bill of Human Rights – approved by the United Nations in Paris on 10 December 1948 – is contained in the Portuguese Constitution. The detailed study of the same comes out overall and in innumerous sections of the "Education Bill".

The first three articles are:

Article 1, number 1, the absolute right to an education is acknowledged. The definition of the educational system is the means of putting this right into effect.

Article 2, number 1, declares that every Portuguese person has the right to education and culture under the terms of the Constitution of the Republic. The Portuguese state therefore accepts responsibility for two of the human rights contained in the Universal Declaration of Human Rights (Articles 26 and 27).

Article 2, number 3, guarantees freedom of choice both from the teaching and learning point of view, tolerating the possible personal preferences. The state may not intervene in the programming of educational courses, stating preferences of certain doctrines - philosophy, politics, ideology or religion. In this way, Articles 18 and 19 of the Bill of Human Rights have been fulfilled.

The direct connection between the main points of the declaration and the law governing the Portuguese educational system is evident.

1.3.4.2 The proposal for overall reform from the Educational Reform Committee

At the moment, Portugal is undergoing its educational reform. This started on 22 January 1986 with the setting up of a committee of eleven carefully chosen people. An overall reform proposal was put forward in June 1988. The active presence of the contents of the Bill of Human Rights is clearly marked in the programme.

The presence of human rights influences a set of basic proposals for the educational reform which are: education for autonomy and freedom; education for democracy; education for development; education for solidarity; education for change.

Three examples referred to, highlighting the defence of human rights and the enormous social problems facing us are: Programme A1 – implementation of a multi-faceted school type; Programme A2 adoption of a new curriculum in primary and secondary schools; Programme B8 – solutions for the underprivileged students.

The idea of a multi-faceted school has been piloted and is slowly in the process of adoption by both state and private schools at all levels (apart from higher education): 20 schools in 1987-88; 44 schools in 1988-89; 68 schools in 1989-90. The results have been very encouraging.

The proposal from the committee is based on the "Education Bill". The law states that both curricular and extra-curricular activities should take place in the schools at all levels, giving a far more complete education to the students. Possible extra activities to be induced have brought a richer type of school and are the model for the future. The traditional school made up only of the curriculum and classes should eventually add extra activities in the form of school clubs to make it more complete in its educational functions.

The proposal goes on to explain that schools need adequate time and structure to adapt from being exclusively academic oriented to being multi-faceted.

Time is measured in schools purely on the basis of lesson time. Each school should be given a specific period of time to organise and develop extra-curricular activities as stated in the activity programme.

The monitoring of time given to schools for extra-curricular activities will go something like this:
a. relating to activities out of the classroom, mainly regarding school clubs, but also projects and activities requiring interaction;
b. back-up to subjects studied, projects and activities related to school subjects;
c. relating to diverse interactive activities chosen by the school committee.
The principal objectives of the "Educational Bill" consists in promoting the following type of extra activities: civic/cultural enrichment; sports and physical education; artistic education; adaption to the community.

The flexibility of this type of school, as shown in practice over the past two years, has been particularly effective with both children and teenagers presenting learning difficulties and those not presenting them. This system promotes a more integrated education with educational programmes and evaluation criteria from a more personal standpoint.

Further on, I shall give some examples of activities carried out by the "Cultural School Project", the one whose objective is a multi-dimensional proposal from the Educational Reform Committee.

The integration of civic education in its entirety in the educational programme from a curricular point of view is clearly focused in the "Education Bill". Article 47, number 2, states: curricular plans for primary education will include all facets – personalised education including ecology, consumer education, family education, sex education, prevention of accidents, health education, educational training for civic services, and so on.

The proposal from the committee includes all aspects of primary education – character development and social training with schools providing the necessary means. The law indicates that while the idea is sound, the integration is not imperative. It was thought that some of the components, or parts of them, could be included in pedagogical activities from other spheres. The Portuguese language is a particular favourite in this sense. One hour a week should be set aside for autonomous development in this area.

In June 1989, the government approved new curricular plans, which particularly coincided with the proposal from the committee by including an optional subject – moral and civic education – in the primary syllabus (the proposal from the committee suggested this should be obligatory), leaving it up to the schools themselves to elaborate the programme.

As a result, students of all nine years of obligatory education have the choice of attending religious education (Catholic) and moral and civic education – na subject more geared to human rights. One of the two is compulsory.

There is the possibility for both primary and secondary students to learn more about civic responsibility as an optional subject.

School is the place to give children an overall picture of human nature – its behaviour patterns and its rights – so we must see this subject as a valuable one. On the one hand, children learn something about human rights, on the other we could fall into the trap of indoctrination, which would theoretically be manipulating and considered an infringement of rights. The truth of the matter is that all activities and school subjects have something to do with her. It is the school committee which either shows a favourable attitude to this, or an unfavourable one. The school, where a child spends so much time, and the teacher, have an enormous influence on concepts of human rights.

Finally we refer to programme B8 which deals with special education. The thinking behind the programme is that it is a sacred human right to be as we are. The Portuguese state declares that it must "provide education respecting each individual unconditionally". Such a stand implies that the educational needs may have to be adapted to suit each and everyone. This is an extremely difficult challenge for the country which has not really got to grips with this problem. The overall proposal from the committee suggest that this problem is now being tackled. "The education system" can only be for everyone if there are no barriers. Directions must be taken and choices must be made.- the best approach would be to try and normalise everything, turning the daily educational life as near as possible to community life for the underprivileged. The disabilities and permanently limited motor functions of the handicapped, especially the serious cases, confront us with a dilemma: either we advocate the right to education in name only, or accept it as a true one, bringing along with it a whole host of practical problems. We, as Portuguese, have to face up to this.

1.3.4.3 The referendum from the Committee for the Protection of Human Rights and Equality in Education

Late in 1988, the Portuguese Ministers of Justice and Education were jointly responsible for forming the Committee for the Protection of Human Rights and Equality in Education. The following were acknowledged: (a) the fundamental importance of human rights in education, underlining the identical rights of all citizens; (b) the relevance of integrated civic education, exploring basic notions of liberty and solidarity; (c) the institutional pressure for permanent programmes in the above-mentioned area.

The objectives of the Committee are enlightening as to the Portuguese Government's interest in solving the problem. They are as follows: (a) promotion of human rights and civic education in schools; (b) development of operations with a view to eradicating priorities in education, more specifically, discrimination relating to sex, social position, economical status or ethnic origin.

IEI is on the list of officially sought out groups, be they state-owned or private, national or foreign, to whose ideas the commission should give priority.

The commission will probably soon have a centre of resources, more specifically, a library and games room to be used to give support to schools and other interested parties to be indicated by the local councils.

1.3.4.4 Samples activities of human rights elaborated by some primary and secondary schools

IEI studied projects which were carried out in schools ranging from the 2nd and 3rd stages of primary to secondary education, in the four existing regional education boards, emphasising human rights.

Some of the results, relating to the Regional Education Boards of Lisbon, the south and the north are available. The results are as follows:

Kind of activities undertaken

– "Education Project for Development": various activities combining the pedagogical aspect of the project with the traditional one. This ended with an end-of-year open session.

– Thematic cycles relating to optional religious education (Catholic).

– Exhibition of work by students, related to the Bill of Human Rights.

– Session to motivate and inform about human rights, highlighting the role of Amnesty International.

– Exhibition of Consumer Tendencies – the rights of the consumer as well as rights of pensioners: healthy environment, information and development.

– Exhibition of European Community institutions, in particular the European Parliament and its role in defending human rights.

– Exhibition of Human Rights.

– Activities in European "school clubs".

– Unesco activities.

– Commemoration of the 40th anniversary of the Declaration of Human Rights.

– Commemoration of the International Day of the Child (Rights of the Child).

– Themes integrated in the Syllabus.

– Sessions organised with the help of Amnesty International.

– Commemoration of the French Revolution's bicentenary.

– Seminar on the Rights of the Child.

– School magazines and newspapers dedicated to human rights.

– "Human Rights in Schools" week (Cultural School Project).

– Themes regularly dealt with in "Cultural School Project".

– "Unicef and Children's Rights" programme elaborated by the "Friends of the School Club", commemorating the 29th anniversary of the Universal Declaration of Children's Rights ("Cultural School Project").

– Exhibition on the International Human Rights Day, assisted by the local branch of Amnesty International.

– Diverse activities focusing on curricular items.

– Activities related to several civic clubs.

We have been unable to obtain information on activities of the Central Board of Regional Education, but we know that several measures have been taken there to promote interest in human rights. Therefore we can conclude that the subject of human rights is pertinent and widely discussed in Portuguese schools.

1.3.4.5 Reintegration of migrant worker's children

One of the most interesting current topics being researched by IEI at the moment is the integration of Portuguese emigrants' children in the Portuguese education system. Tens of thousands of emigrants have returned recently. The integration of their children into Portuguese society and schools presents serious problems. We intend to study this problem in depth as the failure to produce results is very high in this group.

We plan to study the following activities regarding the above-mentioned topic:
a. research the distribution/density of emigrants' children of school age, throughout the schools and years of schooling;
b. research the previous school record when entering Portuguese schools;
c. research the factors leading to bad results;
d. research the necessary conditions for a correct integration into the Portuguese education system;
e. adopt measures deemed necessary for their total integration in schools and society, helping them to overcome difficulties resulting in a cultural change, different system and unfamiliar language;
f. put these measures into force.

The definition of measures to be taken may be concluded by the end of 1989, enabling us to start enforcing them in 1990.

"Cultural School Project", started and supervised by IEI, has had great success with the promotion of human rights. The 44 schools integrated into the "Project" have started 656 school clubs, chosen by the students, with support from the teachers responsible.

School clubs where, directly or indirectly, themes pertinent to the human rights one were: defence of the consumer, community intervention, ecology, interculture, environment, European club, social communication, journalism, radio.

The activities held are mentioned in point 1.3.4.4

The second annual meeting of the Cultural School was held in Lisbon on 26 and 27 June 1989. A display of work and a cultural show took place, organised by the 44 schools which make up the "Project". One could see in both sections a strong presence marked by the awareness of human rights in those schools.

Also in the *Register of Research/Innovation Projects in Education* which was organised and published by IEI in 1988, there is an awareness from the scientific-educational community. Of the 223 registered projects, 6 are related to this area. They are:

Project   9:   Education, Innovation and Development
Project   10:   Values in education
Project   55:   Integrated systems in education and development
Project   75:   Social and cultural values of children from the Vila Real
                      area – at home and in infants' schools (age group: 2-6)
Project 157:   Educating a sense of values in basic education
Project 253:   Ways of making primary education (or otherwise!) in the
                      light of space and socially changing trends.

The first national competition relating to innovation "Educate with New Ideas/Innovating Education" which took place in 1988 had 184 entries. 45 projects are being sponsored by the institute. Some of them are specifically connected to the theme of education for human rights and democratic values.

Five of them are worthy of note:
– Project for students with learning difficulty.
– Put into action participation in education.
– Educate, how? Stimulating, preparing.
– The citizens' school.
– Overcoming fear of taking pleasure in culture.

Finally, we would like to mention the first competition on a national scale of research projects in education. It started in 1989 and had 67 entries, two of which are worthy of a special mention:

- including a civic and social component in the syllabus; for basic education (2nd and 3rd stages).
- Multicultural education in Portugal.

1.3.5 CONCLUSION

As we have tried to show, Portugal has a competent national institute which handles research and innovation in education: the IEI.

The planned activities have had reasonable results even though the institute has only recently been formed.

These initiatives are worthy of note – human rights education and democratic values are all treated in depth. IEI is and institute which we can consider to be a healthy sign in a democratic country like Portugal.

1.4
Psychological Development and Personal and Social Education in Schools

Bartolo Paiva Campos

Portugal

1.4.1 SUMMARY

This paper will review the research work, surveys and experience in Portugal related to the theme of this symposium. Socialisation and education in democratic values and human rights are topical subjects in Portugal where, from the first to the twelfth year of schooling, the curriculum has been extended to cover personal and social education. The emphasis on psychological development can be explained by the fact that this is one of the aspects to which Portuguese authors attach the greatest value in the context of education for democracy.

National surveys on the socialisation for democratic life of young people in Portugal have revealed: passivity when it comes to participation in political life; support for the democratic system, distrust of the way it works; values tending immediate to favour gratification rather than long-term projects; a focus on immediate rather than on the wider social scene. In a similar vein, if some studies are to be believed, the level of psychological development of most young people is not really adequate for spontaneous co-operation and participation.

Although schooling is not the only factor influencing personal and social development, some research (almost always case studies) has underlined the negative role of the hidden curriculum on education for democracy. The methodology in textbooks does not implicit encourage co-operation, participation and responsibility, observed teaching methods tend to lead to dependency and competition, giving pupils neither a say in decisions nor responsibility for managing projects, areas or time; even student unions appear to fail in the civic education of their members, except in the case of their leaders.

Initiatives to make education for democracy part of the curriculum after April 1974 had a fleeting existence in 1975 and 1976. It was only with the advent of the new Education Act in 1986 and the subsequent reform of the curriculum in 1989 that the following possibilities for personal and social education were created in schools: setting up an interdisciplinary field across all subjects of the curriculum; making room for this dimension in the curriculum in the form of space for activities and projects; creating a subject "personal and social education offered as an alternative to "moral education and Catholic religion".

Several local experiments may provide a starting point for the future curriculum development, although there is still a long way to go. Several Portuguese authors have emphasised on the one hand the importance of developing personal values and interpersonal skills rather than acquiring physical skills and knowledge, and on the other that of intervening in the school's ecological system, in addition to more traditional curricular initiatives.

This paper is based on research work, surveys and experiences developed in Portugal about "socialisation and education in democratic values and human rights", the theme of this symposium. The paper's title can be explained by the fact that in the current Education Bill such matters form part of "personal and social education" and also by the fact that the psychological processes involved are one of the aspects which have received most attention in Portugal.

The *first part* will present the *results of socialisation* [1] especially in young people: firstly as regards political participation and attitudes, as well as human values; and secondly the acquisition of those psychological processes which are necessary life's various problems effectively, including those relating to the exercise of democracy and the respect of human rights. The *second part* refers to the *role played by schooling* in personal and social education. The *third part* will be devoted to the *role which schooling ought to play* in such education.

1.4.2 SOCIALISATION LEADING TO PASSIVITY, INDIVIDUALISM AND CONFORMITY IN YOUNG PEOPLE

The results relate to four areas: participation in politics; political attitudes; human values; psychological processes. The survey does not cover knowledge.

1.4.2.1 Poor participation in politics

As regards *voting in elections*, a surveys carried out by ICS [2] showed that the abstention rate of young people was higher than the national average. The same survey showed that barely 20% of respondents claimed to take part in election campaigns and of these, 8% did so only sporadically.

With regard to *participation in and organisation*, only 5% of young people claimed to be active members of a *political party*; this figure included 3% who belonged to the youth wing of a party. However, according to a survey carried out by NORMA [2], only 2.4% of the general population claimed to belong to a political party. The ICS survey showed that 45% of young people belonged to an association, whereas a survey carried out by IED [2] found that only 24.7% of those interviewed did so; it must be noted that the ICS survey covered young people under 30, whereas the IED one covered those aged under 25. A large number of these associations however were devoted to sports or cultural pursuits. If we restrict our analysis to student unions or trade unions, we find enrolment figures of 13% and 17% respectively; barely 16% of respondents claimed to have attended at least one meeting or annual general meeting of a student union, trade union or vocational association (ICS).

Surveys of young people have highlighted the following activities designed to put pressure on leaders and public opinion: attending parliamentary or local authority meetings (5%: IED); submitting protests, petitions or expressions of support to those in power (8.7%: IED; 22%: ICS); taking part in demonstrations (17.8%: IED; 33%: ICS); going on strike (2.1%: IED; 22%: ICS, although only 2% did so frequently); attending meetings in order to solve collective problems (11.7%: IED; 34%: ICS); working with others to solve social problems (12%: IED). The higher percentages found in the ICS survey may partly be explained by the fact that it included older respondents.

This kind of participation is also low among the general population according to NORMA: barely 3% of respondents claimed to work frequently for a political party and 8% did so occasionally. 3.9% took part in demonstrations frequently and 12.2% did so sometimes. 1.9% frequently canvassed for votes and 7% did so occasionally. The sole form of participation which attracted a higher number of respondents was passive, i.e. listening to the news and following discussions on the radio or television (11.8% frequently and 35.1% sometimes).

Braga da Cruz, who directed the ICS survey, (1985) concluded that the factors which apparently watered down young people's participation in politics were no different from those influencing the population as a whole, i.e. the level of political development and political culture. Reis, in the IED survey, defined the profile of young people who in 1985 were readier to participate in politics as follows: male, belonging to an upper, or middle/upper social group, living in an urban area, educated to beyond primary school level and not a Catholic.

1.4.2.2 Political attitudes: a choice in favour of the democratic system but distrust of the way it works

Attitudes are generally seen as major factors influencing behaviour. The attitudes studied concerned national pride, models of social transformation, the preferred political system, the working of the democratic system, interest in politics and political position.

On the subject of *national pride*, according to the IED survey barely 30% were proud of their country. However the MJ [2] survey showed a different picture in that three-quarters of the young respondents claimed to love their country and barely 20% claimed not to care. The elements of a positive national image were the beauty of the landscape, the pleasant climate, the historic heritage and a good reputation for sports. Around 90% of youngsters were agreed on these. 60-70% mentioned conviviality, a democratic regime which respected citizens' fundamental freedoms, investment capacity, ability to work and the same level of technical or scientific skills as was found abroad. The symbols and emblems of the "fatherland" (the flag, the national anthem, parades) are losing ground; "we are witnessing an overall decline in the rituals of nationality which are giving way among young people to a nationalism more pragmatically rooted, closely linked to and stimulated by the country's tangible dynamics, finding in this the foundation of effective national representation vis-à-vis the outside world" (Conde, 1989, p. 39).

The IED survey measured *models of social transformation* in terms of how effective the various forms of political participation were deemed to be when it came to solving problems: youngsters preferred the reformist model, rejecting the revolutionary one, although a fair number defended the conservative model and considered the reformist model to be ineffective. In the FCGM [2] survey, 82.7% of parents and 71.6% of their children agreed with the statement, "Security and justice are to be found not in revolution but in well-directed evolution". However 22.4% of young people and 11.1% of their parents agreed with the statement "In some circumstances violence is the only adequate remedy".

When questioned about the type of *political system* which ought to guide the nation, i.e. a single-party system or a multi-party one, 76.2% of parents and 85.3% of their children (university students) opted for the latter. Similarly, in the NORMA survey, only 12.3% of young people and adults preferred a single party; although 6.1% chose the reply "none", 59.3% opted for having several parties (only two: 20%; three or four: 32.7%; other answers: 6.6%); 19.8% had no opinion.

Although a multi-party system was preferred, the *way it worked* aroused negative attitudes. In the IED survey young people were critical of or even lacked confidence in the two main pillars of the democratic system, political parties and elections, agreeing that parties "divide people more than they unite them, and are dominated by half a dozen individuals whom others follow". As for elections, 49.7% of respondents felt that people were swayed when voting, with 10% even claiming that people voted at random; only 38% felt that people voted consciously. Similarly, the MJ survey reported that about half the young people saw Portugal as a country where party dynamics were heavily involved in the game of power (30% disagreed). It must be noted however that doubts concerned not the lack of efficacy of the political regime per se (41.5% did not think that Portugal had an ineffective political regime and only one third claimed that it had) but rather the poor working of the party system. In the NORMA survey over 60% of young people and adults agreed that "all parties are the same" and that "they only divide people" - although 50% of respondents felt that "there can be no democracy without political parties" and that "people can take part in politics via the parties".

The lack of confidence in the working of the democratic system goes hand in hand with a *lack of interest in politics*. In the IED survey 89% of respondents showed little or no interest in politics. In the ICS survey only 9% claimed to be very interested, 61% were moderately interested and 27% were not interested at all. In the MJ survey social and political participation was the least valued of the six dimensions concerning each respondent's personal future; only 3% classed it among the first three priorities. Its low value can hardly be ascribed to competition from the other five dimensions (vocational fulfilment, money/material security, family life, emotional life, leisure) since 36% of respondents were indifferent to this dimension whereas the figure for indifference in the case of the other five dimensions never exceeded 10%. Even the new social movements such as pacifism and ecology have not managed to capture young people's interest to any greater extent. In the IED survey the numbers belonging to these movements were not very high - although higher for ecology than pacifism.

The results of the various surveys (IED, ICS, FCGM) are broadly similar as regards respondents' political positions: most placed themselves in the centre, with similar numbers belonging to the right and to the left. In the FCGM survey, which allowed comparisons between parents and their children (university students), young people tended to be more to the left and their parents more to the right, although the majority in both generations preferred the centre.

1.4.2.3 Human values: greater priorities given to personal values

At a more general level than for attitudes, values too are often related to behaviour. In the 80s several studies were carried out on the values of Portuguese youth, some of these studies being restricted to university students. One such study is an inter-generation survey which provided data on adults. All studies used the Rokeach scale of values.
The IED survey only dealt with basic values [3]. A first analysis concluded that personal values were more important than social values (Correia Jesuino, 1983). The same author compared these results with those obtained using the same scale on samples of young people in 1980 and 1983; he found that they tended to differ from the 1980 results and tended to agree with the 1983 ones; in 1983 the preponderance of personal values over social values was even more marked.

In a second analysis of the same results Vala (1985) reclassified the values into personal values (subdivided into hedonistic values and values of development, expression and self-affirmation), relational values and social values. He found the following hierarchy, in decreasing order: (a) personal values associated with immediate satisfaction and gratification; (b) relational and affiliative values; (c) personal values of development and achievement; lastly (d) social values. Young people therefore seemed to centre their interest on immediate gratification rather than on long-term projects and to focus on their immediate neighbourhood rather than on the wider social scene. Moreover the author noted that students preferred expressive personal values and social values, whereas young workers gave more weight to hedonistic personal values.

In the FCGM [2] survey, students from a municipality close to Oporto also held relational values to be more important, whilst their parents attached more weight to social values. When it came to instrumental values, although parents' dominant values were ones involving a relationship even if only an internalised one (morality values), their children's dominant values were personal (self-fulfilment values).

For Correia Jesuino (1983) the supremacy of personal values confirmed the enduring predominance of individualism as a trait of Portuguese culture, a trait which has been identified in other studies [4]. The decline in the relative importance of social values as surveys approach the 90s leads him to conclude that young people are moving towards a position where personal values carry greater weight than social values. This conclusion led Vala (1986) to wonder whether a neo-individualism was emerging. "In addition to what would appear to be the emergence of a forceful search for personal autonomy, might it also be possible to note increasing feelings of helplessness and lack of control over social and political events? If this is so, we can understand that the values of equality, solidarity and social intervention are losing ground whereas individual strategies for solving real-life problems are turning into values" (p. 26).

1.4.2.4 Individualistic and conformist psychological processes

Today it is no longer thought that psychological development is purely the result of maturation. Increasingly, reference is made to the effects of socialisation, especially at school. Unfortunately, there are not many studies in Portugal on the level of psychological development of youngsters on the threshold of adult life and those which do exist are not representative of the whole population.

A study on *strategies of interpersonal negotiation* carried out with a sample of 18-year olds attending to a secondary school in Oporto found that the level reached by almost all the respondents was one where all strategies used in an interpersonal relationship were designed solely to meet personal interests, despite recognition of the interests of the other. There was no recourse to strategies of co-operation and mutual satisfaction of interests (Coimbra and Campos, 1989).

A study carried out with a representative sample of 2nd-year students attending five university courses in Oporto examined the process undergone in defining their *identity*, i.e. the complex of goals, values and beliefs which goals give direction and meaning to people's lives (vocational, interpersonal, political goals, etc.) (Costa and Campos, 1986;

1989a, 1989b). The following results were obtained in respect of four different types of process defining their identity: (a) in about 40% of cases what was observed was not a personal commitment resulting from prior exploration of possible alternatives but a commitment corresponding to acceptance without question of the choices and projects of significant persons or of authority; (b) 8% were in an undefined situation, characterised by a lack of commitment; (c) 20% were in the period of asking questions and exploring alternatives; (d) about 30% had already made their personal commitment. When the same respondents were observed three years later at the end of their studies the figures were (a) 30%, (b) 8%, (c) 22% and (d) 42% respectively. The proportion of autonomous identities had increased slightly whereas the figures for foreclosed and conformists had fallen. The quality of this development was not even throughout the different university courses.

When the same study analysed *political* identity, the figures for the various categories were (a) 27%, (b) 52%, (c) 14% and (d) 8% respectively the first time and (a) 30%, (b) 28%, (c) 30.5% and (d) 11.5% three years later. In fact politics had no meaning in the life of half the 2nd-year students. By the end of the course this still held true for 30% of students but already 30.5% were asking questions and exploring alternatives. Both times, personal commitment was only seen in 10% of cases whereas the figure for conformist commitment was 30%.

The results of this research provide clues as to why the majority of young people show an insufficient level of development in some of the psychological processes which are important for co-operation, personal responsibility and autonomous participation. However, further research is necessary.

1.4.3 WHAT SCHOOLS HAVE BEEN DOING

Even though political behaviour and attitudes are closely related to education for democracy values and psychological processes are also an important part of personal and social education. It is clear that schooling, and particularly length schooling, is a major factor. No studies are available in Portugal to demonstrate the specific influence of schooling; however there is available a series of analyses which suggest education's lack of concern with these question and even that it may have a negative role in the development of attitudes.

Leaving aside studies of the period under the dictatorship (Bivar, 1971; Cortesão, 1982; Formosinho, 1987; Mónica, 1978) we shall examine those covering the period following the setting up of a multiparty democratic system in 1974. These studies refer to the hidden curriculum (teaching methods, the structures and organisation of schools) and to specific initiatives introducing education for democracy into the curriculum, including personal and social education.

1.4.3.1 The hidden curriculum

Various studies, which have almost always been exploratory and qualitative in nature, covering a small number of cases, have attempted to reveal certain elements of schools' hidden curriculum which make it difficult to pursue the objectives of personal and social development, especially education for democracy.

Valente (1988) co-ordinated the analysis of the text-books on natural sciences, social studies, the mother tongue and mathematics used during the second half of the 80s for the first six years of schooling (which was the basic education up until 1987). This analysis, which dealt with the values presented in the textbooks, was almost always critical. We shall take as an example the analysis of the values of co-operation, participation and responsibility found in the textbooks on natural sciences. In all six textbooks used there were only four suggestions for group work whereas the number of individual activities proposed was 124. "We can state that the motivation to carry out group work is poor in the textbooks and that the co-operative organisation of schoolwork is suggested even less often. It is very rarely suggested that projects involving inter-class work or *co-operation* are necessary within the school, or even those involving inter-school co-operation by exchanging materials and/or correspondence, and the community is considered as a useful resource only sporadically" (p. 194). On the subject of *participation*, the study concluded, "This is not one of the authors' concerns (...). Active participation by pupils is extremely low, as regards their own development, the school or the environment. The pupil can neither choose his work nor help organise it" (p. 197). Nor were there any proposals for activities which would specifically encourage *responsibility*: "The activities proposed are over-structured, offering pupils few opportunities for organising work in the light of their experience, interests and inclinations" (p. 198).

An analysis by Marques (1989) of the textbooks for the first nine years of schooling concluded that, although the syllabus laid emphasis on values such as justice, human dignity, reciprocity, tolerance, the right to life, a critical approach and civic rights, most textbooks were far from granting them the same status. But around 1985 an analysis of syllabuses for the first six years of schooling found that intellectual development was clearly at a premium whereas physical, emotional, aesthetic, social, moral and spiritual development was ranked lower (Soares and Abreu, 1986, p. 103) (7).

Other studies dealt not with the teaching-learning process proposed in textbooks but with teaching methods and the organisation of schools.

Bettencourt and Brederode Santos (1981) drew up a list of objectives and skills (individual, interpersonal, social and suprasocial skills) which needed to be enhanced in personal and social education; they carried out an exploratory study to determine whether *the teaching methods and organisational conditions* in some primary and secondary schools in Lisbon did, in fact, contribute to the enhancement and pursuit of these objectives. They concluded that, although syllabuses in the late 70s affirmed a good number of these objectives, teaching methods and the organisation of schools worked against these. For example, the syllabus affirmed the importance of individual activity in forming a group and in developing a spirit of co-operation and positive habits of relationships and friendship. "In practice, however, teaching — which continues to be essentially centred around the teacher — inevitably constitutes training in dependency. Group work is more widely used than hitherto but teachers do not always have the training to support it, and the school's facilities do not always favour it. Moreover, overpopulation in schools, inadequate material conditions and at times the lack of explicit rules governing collective life encourage the apprenticeship of irresponsibility, violence, oppression and a struggle for survival, far more than they promote responsibility and co-operation " (p. 45).

In an intensive study, Marques (1989) found that the results obtained in primary and secondary schools confirmed the hypothesis that schools' hidden curriculum took no account of the manifestation of values which were the expression of pupils' moral development. "They present a spatial organisation and a relationship between pupils, teachers, parents and management which gives no opportunity for participation in decisions." The author listed a series of aspects of the hidden curriculum which hampered the pursuit of objectives in the social and moral field. These included: the absence of an area and time under pupils' control, intended for collective participation and for solving problems affecting individuals, groups and the school; the mobility of the teaching profession, making it more difficult to keep pace with and know pupils; overpopulation in schools, preventing personal relationship with teachers.

Bettencourt and Marques (1987) carried out a qualitative study of two primary and two secondary schools. On the school's role in developing pupils' democratic values, they concluded, "It generally restricts itself to transmitting academic knowledge, leaving learning about and exercising the rights and duties of citizens to the family and the mass media, particularly television and radio" (p. 106). According to these authors, schools failed in respect of pupils' moral and civic development because, "It gives them no responsibility in managing projects, areas and time. Without actual participation, pupils lack the experience of exchanging opinions and confronting points of view, ideas and values, these activities being essential if they are to construct a code of conduct distinguished by autonomy, a critical approach and the ability to take decisions, to accept different points of view and to change opinions or attitudes" (p. 107).

Lima (1988) analysed *pupils' role in managing* secondary schools, which was made possible by the standards for school management introduced after April 1974. He concluded, "Pupil participation in management does not seem to be part of the pupils' horizons and concerns" (p. 152) — this situation being reinforced by the organisation within schools. Brederode Santos and Roldão (1986) note that the management of schools has been more management by teachers than by pupils; it has never been considered a form of civic education for the pupils.

The role of secondary school *student unions* in promoting the social and moral development of young people and in their civic education was the subject of a case study carried out by Brederode Santos and Roldão (1986). The results of this study contradicted the authors' initial positive expectations of students unions. Both pupils and teachers viewed student unions as bodies offering services and organising meetings, sports events or, to a lesser extent, cultural events. Pupils played a very minor role in decisions concerning school life, this being criticised by older pupils. Apart from the leadership of the unions or those directly connected with the leadership, other pupils criticised the way unions were run for their lack of democracy. For the former, playing an active role in the unions enabled them to develop feelings of solidarity, responsible attitudes and social skills; for most pupils the union was merely an opportunity for meeting people and making contacts.

1.4.3.2 Specific curricular initiatives

The specific initiatives placing education for democracy on the curriculum in Portugal after April 1974 followed different strategies between 1974 and 1976:

a *concepts* concerning democratic institutions were *covered* under "physical and social environment" in the first 4 years of schooling and "social studies" in the 5th and 6th years;

b in the 7th school year, *room was made in the curriculum for cross-curricular* projects, linking with the community, called "civic and polytechnic education"; a year of civic service was created prior to university entrance, to be carried out in the community;

c the *subject* "introducing to politics" was added in the 10th and 11th years of schooling.

According to Brederode Santos (1984, 1987), both the objectives and methodology of "introduction to politics" and concepts inclusion within other subjects form part of the movement in education for democracy which stresses *information* to be acquired, and eventually the critical appraisal of such information. For their part, "civic and polytechnic education" and civic service belong to the movement which stresses the action to acquire the *skills* necessary for democratic practices. She found that neither the initiatives belonging to the movement which stresses the development of moral reasoning, nor those belonging to a wider movement of education for development had yet reached Portugal.

Almost all the above initiatives in the field of education for democracy were suspended in 1976 [5]. They could be criticised on two fronts: the way they were set up and the way they were conceived. Bettencourt (1982) studied civic and polytechnic education and Bettencourt and Brederode Santos (1983) studied civic services. On the one hand they were improvised experiences which had been agreed upon only a few months before they were set up, they received little or no support and the teachers were inadequately prepared. On the other hand it was claimed that they took no account of the chief target of these innovations, i.e. pupils. They overestimated pupils' true capacity to learn bearing in mind their age and stage of development. Despite these reservations, both authors acknowledged the positive side of the experiences and their potential in other, future circumstances. According to Brederode santos (1984, 1985) the criticism underpinning the decision to suspend the experiments – i.e. risk of introduction – was not justified. She acknowledged however that the driving force behind this experiment was "an attempt to promote more just forms of society without at the same time promoting the maximum development of the individuals making up such societies and justifying their existence" (Brederode Santos, 1987, p. 131) – a dangerous attempt. This author added that it was no less dangerous "to take no account of the school's function of indoctrination and its role in pupils' social and moral development", since it can continue to promote the values of the "old school" via its hidden curriculum, whilst flying the flag of neutrality.

1.4.4 WHAT SCHOOLS CAN AND MUST DO

1.4.4.1 Under the new Education Act

The new Education Act governing the Portuguese educational system (Act 46/86 of 14 October) places squarely upon schools the responsibility for education for democracy and human rights. Schools must help form citizens who are free, responsible and autonomous and who show solidarity. They should develop the democratic, pluralist spirit which respects others and their ideas, and is open to dialogue and the free exchange of opinions. They should form citizens who are capable of judging their community with a critical, creative eye and are capable of commitment to its gradual transformation. They should allow experiences which promote civic awareness and emotional maturity, create attitudes and practical habits of co-operation within the family or by intervention in the community. They should promote national feeling which is open to reality in a spirit of universal humanism, solidarity and international co-operation (Articles 2 and 7).

To reach these objectives, the Act states that the basic education curricula shall "include at all levels an element of personal and social education, in an adequate form, possibly comprising ecological, consumer, family and sex education, and civic education (Article 47 [2])

1.4.4.2 Reform of the basic and secondary education

At the beginning of August 1989 the government took the following decisions in respect of personal and social education in schools, having heard proposals of the Commission for the Reform of the Educational System (CRES, 1988) and the opinion of the National Council on Education (NCE, 1989; Campos, 1989a):

a Personal and social education shall be a cross-curricular theme; in other words all the components of the curriculum of basic and secondary education shall systematically contribute to pupils' social and personal education, promoting the acquisition and internalisation of spiritual, aesthetic, moral and civic values.

b Personal and social education shall be one of the three major objectives (the others being the concretisation of knowledge and the links between school and the community) of an *area in the curriculum* for activities and projects the organisation of which shall be left to the school within the timetable for the subjects involved in each project; in the 7th, 8th and 9th years of schooling personal and social education shall be called "civic education", with a national syllabus covering the democratic institutions. Pupils will be assessed.

c Personal and social education shall be placed on *the curriculum as a subject* (one hour per week) under the title of "personal and social development" in all years of primary and secondary education, *but only as an alternative* to the subject "moral education and the Catholic religion" (or other religions); the Catholic Church proposed that pupils doing "moral education and the Catholic religion" should be released from doing "personal and social development" — a proposal accepted by the government (Campos, 1989a).

1.4.4.3 Development of the personality: from moral reasoning and the clarification of values to personal development

Setting up "personal and social development" constitutes a major challenge to teachers, schools, training and research centres and the Ministry. In 1988 the Ministry set up a Commission for Promotion Human Rights and Equality in Education, which will doubtless tackle this task. The proposals put forward by some authors, usually inspired by foreign authors or projects and a few local experiences will no doubt provide a starting point for taking action — although there is still a long way to go. Let us first review the proposals made.

A project commissioned by the Ministry of Education (CEDCEP, 1988) highlighted social and moral values as one of the three important dimensions on the profile of a pupil having completed his secondary education. Beyond listing the general objectives of such education, for each objective the project specified behaviour and attitudes expressing it and proposed appropriate teaching strategies for promoting it. Most of these were related to the hidden curriculum.

However, the majority of Portuguese authors have shown themselves to be concerned most with the psychological processes inherent in personal and social education in general and in education for democracy in particular.

Brederode Santos (1984, 1985 and 1987), who has taken the greatest interest in this question and has adopted the most systematic approach, has stressed the low status of such education in Portugal. Education in values seems to be merely the inculcation of social values. She has presented both Kohlberg's North American movement for moral education (ranging from the discussion of dilemmas to the democratic organisation of schools and including specific psychological education programmes) and Raths and Simon's clarification of values, centred around the process constructing values, not the values per se. Nor has she omitted to mention Ryan's claims that education in the actual content of values which represent a consensus in a given society at a given time, is necessary.

Formosinho (1986), also adopting Kohlberg's view, has analysed the various strategies of socio-mcral education which she has proposed for Portuguese secondary schools. Marques (1986) has even proposed that value education be introduced in preschool education and has set out a methodology for this age group, in line with Kohlberg's model.

Having emphasised that educational practices inevitably influences values, Valente (in the press) defends the deliberate use of the school's role in value education. She lists the various doubts and fears raised by such a project: the cultural stereotype that it is a private matter; the fear of indoctrination; distracting the teacher's attention from the imparting of knowledge and skills; the idea that the appropriate place to deal with these matters is the family, the church or the party; lastly, the fear that teachers are inadequately prepared for such a delicate task. She presents two approaches as alternatives to the traditional inculcation of value: Raths' clarification of values and Kohlberg's socio-moral education, mentioned above. Stating that neither approach is designed to inculcate pre-established models but to develop processes, the author ends by referring to personal and social education., highlighting the fact that this is not limited to information about contents and emphasising certain psychological processes and skills which have to be developed in order to solve problems in the various fields covered by this education: socio-moral reasoning and clarification, interpersonal relationships, motivation, autonomy and identity. She proposes a specific curricular field and the contribution of all subjects, drawing attention to the importance of the school's atmosphere.

Most Portuguese authors have stressed the need to take into account the *whole complex of psychological processes* (not merely moral reasoning and clarification of values) which need to be developed if people are to acquire skills to solve the various situations they encounter throughout life flexibility and creatively. These include field of civic participation or in other fields such as family, sexual, vocational or economic matters.

Campos(1980, 1985, 1988) and Brederode Santos (1984) have emphasised this approach, which was also proposed by Júlia Formosinho to the Commission for the Reform of the Educational System. The Commission has established the following objectives of personal and social education in respect of psychological processes:

a a broader frame of reference which takes account of all variables and their relationships, works out alternatives and anticipates consequences;

b the ability to understand various points of view (those of others who may be close or distant) and to incorporate such points of view in dialogue and decisions;

c a capacity for empathy, integrating into one's attitudes an awareness of ever-widening circles of human relationships;

d the development of *self-identity* as to include many possibilities, approaches and feelings;

e lastly, the construction of universal values which guide a person's thought and moral position beyond mere conventions" (CRES, 1988, p. 123).

Brederode Santos (1977) has even suggested that the education of citizens for democracy is an education whose aim is the integral development of each individual.

Another aspect emphasised by Portuguese authors taking into account the influence of the hidden curriculum has been the need to act on teaching strategies and the organisation and structure of schools. Campos (1988, 1989b) dealt with the psychological dimension of personal and social education in terms of *intervening in the school's ecological system* in such a manner that the psychological processes to be developed are no longer limited to internal processes but also covered interpersonal processes within groups, organisations and relationships networks. He concludes that the curricular innovation compatible with such an approach is that which gives priority to the restructuring and reorganisation of each school and to reform at local level, albeit supported from outside.

Such a psychological approach can be extended to teachers' professional development. The Education Act states that teacher training should provide an opportunity for the social and personal education which is necessary to exercise the profession (Article 3 (1a)). This training is particularly necessary if teachers are to play a role in pupils' personal and social education. In this sense some training centres have been organising psychological education for teachers (Brederode Santos, 1985; Campos, 1983b, 1983c; Formosinho & Santos, 1983; Conçalves & Cruz, 1985; Joyce-Moniz, 1988.)

The general context is similar to that proposed for pupils, adapted to suit the age group. Beyond the curricular aspects of training, various authors have stressed the influence of the educational context. Brederode Santos (1984) concluded her study by listing the essential elements of a context for teacher training which promoted their development. Campos (1986) stressed that in-service training was not merely a matter of helping teachers to develop psychologically but of acting on the system of professional relationships. Along with others he has proposed a methodology based on taking *action* in a real situation and by *reflection* integrate the outcome in the framework of a collaborative, consultative *relationship* to give the action meaning.

1.4.4.4 Some experiences

In addition to the proposals already mentioned, there have also been a few local experiences in personal and social education, not to mention the national ones set up in 1975 and 1976 which have been covered above.

One such experiment, carried out between 1985 and 1987, centred around the introduction of alternative forms of organisation in three schools (primary, middle, secondary). It concluded that its impact was limited because of the small number of changes made and their limited scope. However, it listed other structural and functional aspects which were urgently in need of transformation (Correia, 1988; Veloso & Rodrigues, 1987). Experimental *cross-curricular project* involving relationships with the community in some Portuguese secondary schools have already been presented at a previous Council of Europe seminar (Starkey, 1987, 9-14). An experiment covering 39 classes and six teachers drawn from five schools of the "Casa Pia" in Lisbon started up last school year; it set out to promote personal and social education by creating a *specific area in the curriculum* devoted to seminars, discussions, study visits, clubs, etc (Freitas Dinis, 1988).

The action of many psychologists who have been working in secondary schools since 1983 is designed to develop the psychological processes underpinning the life-skills necessary for problem-solving throughout life. Despite the obstacles due to their precarious professional situation and to the fact that psychological practice in Portugal is in its infancy, they have attempted to contribute to personal and social education in the schools doing group work with the pupils and by working with individual teachers or groups of teachers, supporting their efforts to transform their teaching strategies, the school's organisation or the school's network of relationships with the community. They do not practice their profession in the abstract, but help teachers confronted with situations of professional concern, pupils confronted with personal concerns: school achievements; vocational planning and decisions; interpersonal relationships (camaraderie, friendship or intimacy); social participation; independence;

autonomy, etc. They do not aim to develop personal psychological processes alone but also interpersonal one (Abreu *et al*, 1987; Coimbra *et al*, 1986; Imaginário & Campos, 1987; Soares, 1987; Soares & Campos, 1986). The Education Act itself emphasises the role of psychological and vocational guidance services which were to be set up in schools with the aim of supporting pupils' psychological development and the system of relationships within the school community (Article 26).

1.4.5 CONCLUSION

Although it is fair to say that over the last fifteen years Portugal has seen a surge of interest in research into education for democracy or personal and social education, and that a good number of research projects have actually been carried out, it cannot be said that this research has enjoyed much priority. Perhaps the main reason has been lack of tradition in the country of research in social sciences, particularly educational sciences. Moreover, such research is not yet deemed to be sufficiently important for the country's development, an economist's vision of development having prevailed to date. For a country which is facing far-reaching transformations in its socio-economic structure as a result of joining the EEC, this can have serious consequences. In this context the development of research in social sciences might prove one of the pillars consolidating and strengthening democracy in Portugal.

Although it seems vital to create educational opportunities at school designed to fit young people for personal fulfilment and social life in democracy, we should not expect too much of this, especially if nothing else is done. Apart from schooling, we have to take into account the influence of other contexts of socialisation and of social values, standards, structures and practices - which also exist at school. The most effective strategy therefore, in addition to a long enough period of successful schooling, might be to aim to transform the social values, standards, structures and practices existing in school life itself. This means tackling educational reform, innovation in teaching and curricular development in a totally different way. It is not the province of the central bodies to produce a reform which schools would then buy off the shelf but to create the conditions for encouraging the development within schools of permanent innovation, with the support of outside centres for training and research. This approach is also valid for developing curricular strategies: spreading personal and social education across other subjects and making it an area or a subject in its own right. Although these curricular strategies have a more limited impact they can make a useful contribution, especially if: (a) they move from the bookish, passive, traditional ways of teaching to become based on meaningful projects representing a challenge for the young people involved; (b) they give each individual the opportunity to integrate personal experience failing which they may be accused of inconsequential activism; (c) they are not limited to acquiring knowledge and modelling values, standards and behaviour but are designed to develop psychological processes and the life skills of people and social groups; (d) they fit the context of personal relationships, the true setting for personal and social development.

Considering that personal and social education is not just an academic subject at national level, and recognising that innovations limited to this form of education may fail in the medium term, there is perhaps an opportunity to experiment another kind of teaching and another model of local curricular development, a model in which teachers, among others, would cease to act as *public servants* and act as *professionals* in the field of education, which is also a matter of democracy.

NOTES

1. Taken here in the meaning of Hurrelman's definition (1988):
   "Socialisation is the process of the emergence, formation and development of the human personality in dependence on and in interaction with the human organism, on the one hand, and the social and ecological living conditions that exist at a given time within the historical development of the society, on the other. Socialisation designates the process in the course of which human being, with his or her specific biological and psychological disposition, becomes a socially competent person, endowed with the abilities and capacities for effective action within the larger society and the various segments of society, and dynamically maintains this status throughout the course of his or her life".

2. Several surveys representative of various sectors of the Portuguese population were carried out in the 80's. For the sake of simplicity, they have been identified by initials:

   IED
   A survey carried out in 1983 by the Institute of Development Studies, with a representative sample of Portuguese youth (Y=1095), covering various questions relating to young people (Correia Jesuino 1983; Reis, 1985 and 1986; Vala, 1985 and 1986).

FCGM
A survey on values carried out by the Calouste Gulbenkian Foundation in and around 1984, with representative sample (N=162) of the population aged 12-25 of a municipality in the North of Portugal (Matosinhos), who were attending school or university. The survey also included their parents (N=296) (Figueiredo, 1985).

FCGM
A survey carried out in 1986 by the Calouste Gulbenkian Foundation with a national sample of university students (N-404) (Figueiredo, 1988).

ICS
A survey on political participation and attitudes carried out by the Institute of Social Science of Lisbon University in and around 1984, with a representative national sample (N=900) of young people aged 15-29 (Braga da Cruz, 1985).

MJ
A survey commissioned from the Institute of Social Science by the Ministry of Youth, carried out in 1986 and 1987 with a representative national sample, (N=1963) of young people aged 15-29 (Conde, 1989, Ferreira, 1989).

NORMA
A survey carried out in 1985 by the market research company NORMA with a representative national sample (N-2000) of young people and adults (Stock, 1988).

3. Final values concern *results*; instrumental values concern ways and means.

4. Other studies on the presence of sex stereotypes in school textbooks or in other school practices have been carried out over the last two decades (Bivar, 1971; Brandão, 1979; Fernandes, 1988; Fontaine, 1978; Leal, 1979).

5. Paradoxically by a government drawn exclusively from the ranks of the socialist party. Since then, the Minister of Education, J A Seabra, has merely appointed, in 1984, a working group to study the introduction of "civic education" as a subject after the 4th year of schooling. This was a political gesture without any practical outcome to date.

1.4.6 REFERENCES

ABREU, M.V. (1985). Orientação escolar e profissional e desenvolvimento da personalidade. Cadernos de Consulta Psicológica, I, 95-101.

ABREU, M.V. (1988). Para uma psicopedagogia do sucesso escolar. In C.R.S.E., Medidas que Promovam o Sucesso Educativo. Lisboa: GEP do Ministério da Educação.

BETTENCOURT, A.M. (1982). La liaison École-Milieu-Production à L'École Secondaire Portugaise. Doctoral dissertation, University of Paris V.

BETTENCOURT, A.M. & Brederode Santos M. E. (1981). Papel da escola na formação democrática dos alunos. In Politica Educacional num Contexto de Crise e Tranformação Social. Lisboa: Moraes Editora/I.E.D.

BETTENCOURT, A.M. & Brederode Santos M.E. (1983) O Serviço cívico estudantil: proposta para uma didcussão. O Professor no. 51, 4-20

BETTENCOURT, A.M. & Marques, R (eds.) (1987). Percursos Escolares, Estratégias de Vida, Códigos de Conduta. Lisboa: GEP do Ministério da Educação.

BIVAR, M.F. (1971). Ensino Primário e Ideologia. Lisboa: Seara Nova.

BRAGA DA CRUZ, M. (1985). A participação política da juventude em Portugal. Análise Social, XXI (87,88,89), 1057-1088.

BRANDÃO, E. (1979). Os Estereotipos nos Manuais Escolares. Lisboa, Comissão da Condição Feminina.

BREDERODE SANTOS, M.E. (1984). Education for Democracy: A Developmental Approach to Teacher Education. (Master thesis, Boston University).

BREDERODE SANTOS, M.E. (1985). Os aprendizes de Pigmaleáo. Lisboa: I.E.D.

BREDERODE SANTOS, M.E. & Roldão, M.C. (1986). As associações de estudantes no ensino secundário. Um modo de promover o desenvolvimento sócio-moral e a formação cívica dos jovens? Desenvolvimento, no. spécial, 79-92.

BREDERODE SANTOS, M.E. (1987). Educar para democracia. in Bettencourt, A.M. & Marques, R (eds.). Percursos Escolares, Estratégies de Vida, Códigos de Conduta. Lisboa: GEP do Ministério da Educação.

CAMPOS, B.P. (1980). A orientação vocacional numa perspectiva de intervenção no desenvolvimento psicológico. Revista Portuguesa de Pedagogia, XIV, 195-230.

CAMPOS, B.P. (ed.) (1983a). Actas do Primeiro Encontro Nacional de Formação Psicológica de professores dos Ensino Preparatório e Secundário. Porto: Serviço de Consulta Psicológica e Orientação Vocacional.

CAMPOS, B.P. (1983b). A formação psicológica dos professores para o desenvolvimento humano dos alunos. Actas do Primeiro Encontro Nacional de Formação Psicológica de Professores dos Ensinos preparatório e Secundário. Porto: Serviço de Consulta Psicológica e Orientação Vocacional, 6-12.

CAMPOS, B.P. (1983c). Que psicologia para professores? Actas do Primeiro Encontro Nacional de Formação Psicológica de Professores dos Ensinos Preparatória e Secundario. Porto: Serviço de Consulta Psicológica e Orientação Vocacional, 88-93.

CAMPOS, B.P. (1985). Consulta psicológica e projectos de desenvolvimento humano. Cadernos de Consulta Psicológica, 1, 5-9.

CAMPOS, B.P. (1986). Formação participante de profissionais do desenvolvimento humano. Revista de Psicologíca e de Ciências da Eduçăo, 1, 7-16.

CAMPOS, B.P (1988). Consulta Psicológica e desenvolvimento humano. Cadernos de Consulta Psicológica, 4, 5-13.

CAMPOS, B.P. (1989a). Formação pessoal e social e desenvolvimento Psicológica dos alunos. Cadernos de Consulta Psicologíca, 5.

CAMPOS, B.P. (1989b). Intervenção ecológica par o desenvolvimento da criança. In Silva, J.L. & Miranda, G. Projecto Alcácer. Lisboa: Fundação Calouste Gulbenkian (in press).

CAMPOS, B.P. et al. (1985). Consulta Psicológica na orientação escolar e profissional em escolas secundárias da região norte. Cadernos de Consulta Psicológica, 1, 139-150.

CARVALHO et al. (1987). Desenvolvimento de competências de estudo nos jovens. Cadernos de Consulta Psicológica, 3, 89-93.

C.E.D.C.E.P. (1988). Perfil Cultural Desejável do Diplomado do Ensino Secundário. Lisboa: GEP do Ministério de Educação.

COIMBRA et al. (1986). Consulta Psicológica e desenvolvimento interpessoal de jovens. Cadernos de Consulta Psicológica, 2, 59-69.

COIMBRA, J. & Campos, B.P. (1989). The relationship between the interpersonal understanding and interpersonal negotiation strategies in adolescents: a cross sectional study. Paper presented at the "Tenth Biennal Meeting of the I.S.S.B.D.", Finland, Juväskylä, July, 1989.

CONDE, I. (1989). A Identidade Social e Nacional dos Jovens. Lisboa: Instituto da juventude e Instituto da ciências cociais.

CORREIA, Jesuino, J. (1983). Valores finais da juventude portuguesa em 1983. In Situação, Problemas e Perspectivas da Juventude em Portugal. Vol. VIII. Lisboa: I.E.D., 121-139.

CORREIA, M.A.P. (1988). A Escola e o Desenvolvimento dos Alunos. A Escola Serve para Educar? Lisboa: GEP do Ministério da Educação.

CARTESÃO, L. (1982). Escola, Sociedade. Que Relação? Porto: Ed. Afrontamento.

COSTA, M.E. & Campos, B.P. (1986). Identidade de estudantes universitários. Cadernos de Consulta Psicológica, 2, 5-11.

COSTA, M.E. & Campos, B.P. (1989a). Social context and identity development. The case of university area of study. In B. Campos (ed) Interpersonal and Identity Development. New Research Directions. Porto: Serviço de Consulta Psicológica e Orientação Vocacional (in press).

COSTA, M.E. & Campos, B.P. (1989b). University area of study and identity development: a longitudinal study. Paper presented at the "Tenth Biennal Meeting of the I.S.S.B.D.", Finland, Jyvaskyla, July.

C.R.S.E. (1988). Proposta Global de Reforma. Lisboa: GEP do Ministério de Educação.

FERNANDES, J.V. (1988). A Escola e Desigualdade Sexual. Lisboa: Livros Horizonte.

FERREIRA, P.M. (1989). Os Jovens e o Futuro. Expectativas e Aspirações. Lisboa: Instituto da juventude e Instituto das Ciências Sociais.

FIGUEIREDO, E. (1985). Mudança, valores e conflicto de gerações em Portugal. Análise Social, XXI, (87-88-90), 1005-1020.

FIGUEIREDO, E. (1988). Conflito de Gerações. Conflito de Valores. Lisboa: Fundação Calouste Gulbenkian.

FONTAINE, A.M. (1978). A discriminação sexual dos papéis sociais nos manuais portugueses de aprendizagem da leitura. Revista Portuguesa de Pedagogia, XI, 149-183.

FORMOSINHO, João, (1987). Educating for passivity - a study of Portuguese Education, 1926-1968. Doctoral dissertation, London University.

FORMOSINHO, Júlia & Santos, M.R. (1983). Reflexão sobre o lugar da psicologia na formação de professores. Actas do Pimeiro Encontro Nacional de Formação Psicológica de Professores do Ensino Preparatório e Secundário. Porto: Serviço de Consulta Psicológica e Orientação Vocacional, 62-68.

FORMOSINHO, Júlia (1986). A intervenção da escola no desenvolvimento sócio-moral. Desenvolvimento, 3, 61-74.

FORMOSINHO, Júlia (1987). Fundamentos psicológicos para um modelo desenvolvimentista de formação de professores. Psicologia, V, 3, 247-252.

FREITAS DINIS, J.F. (1988). Educação ética e social. Revista da Casa Pia, 1, no 2, 12-13.

GONÇALVES, O., SOARES, I. LEMOS, M. & POVOAS, I. A formação psicológica dos professores nos ramos educacionais da Faculdade de Ciências do Porto. Actas do Primeiro Encontro Nacional de Formação Psicológica de Professores dos Ensino Preparatório e Secundário. Porto: Serviço de Consulta Psicológica e Orientação Vocacional, 30-43.

GONÇALVES, O. & CRUZ, J.F. (1985). Desenvolvimento interpessoal e formação de professores. In J. Cruz et al (eds.). Intervenção Psicológica na Educação. Porto: Associação Portuguesa de Licenciados em Psicologia, 199-217.

HURRELMAN, R. (1988). Social Structure and Personality Development. Cambridge: Cambridge University Press.

IMAGINARIO, L. & CAMPOS, B.P. (1987). Consulta psicológica vocacional em contexto escolar. Cadernos de Consulta Psicológica, 3, 107-113.

JOYCE-MONIZ, L. (1988). Formação de professores, desenvolvimento dialéctico e dialécuca no ensino. Revista Portuguesa de Educação, 1, 5-20.

LEAL, I. (1979). A imagem feminina nos manuais escolares. Cadernos da Comissão da Condição Feminina, no 11.

LIMA, L. (1988). Gestão das Escolas Secundárias. A Participação dos Alunos. Lisboa: Livros Horizonte.

MARQUES, R. (1986). A Criança na Pré-Escola. Lisboa: Livros Horinzonte.

MARQUES, R. (1989). A educação para os valores no ensino básico: resultados parciais de uma análise dos curriculos (texto polycopié).

MONICA, M.F. (1978). Educação e Sociedade no Portugal de Salazar. Lisboa: Presença.

N.C.E. (1989). Novos Planos Curriculares dos Ensino Básico e Secundário.

REIS, M.L.B. (1985). Inserção e Participação Social dos Jovens. Lisboa. I.E.D.

REIS, M.L.B. (1986). Tendências recentes da atitude dos jovens portugueses face à politica: análise comparativa com os indicadores europeus. Desenvolvimento, no especial, 67-68.

SOARES, A. & Abreu (eds.) (1986). Progamas. Análise de Situação. Lisboa: GEP do Ministério da Educação.

SOARES, I. & Campos, B.P. (1986) Educação sexual e desenvolvimento psicossexual. Cadernos de Consulta Psicológica, 2, 71-79.

SOARES, I. (1987). Consulta Psicológica e realização escolar. Cadernos de Consulta Psicológica, 3, 81-88.

STARKEY, H. (1987). "L'enseignement et l'etude des droits de l'homme dans les écoles secondaires". Rapport du Séminaire Européen des Enseignants, Carcavelos, Portugal, 9-12 Décembre 1986. (doc. DECS/EGT (87) 20), Conseil de L'Europe.

STOCK, M.J. (1988). A imagem dos partidos e a consolidação democrática em Portugal. Resultados dum inquérito. Análise Social, XXIV, 151-161.

VALA, J. (1985). Representações Sociais dos Jovens: Valores, Identidade e Imagens da Sociedade Portuguesa. Lisboa; I.E.D.

VALA, J. (1986). Identidade e valores da juventude portuguesa-uma abordagem exploratória. Desenvolvimento, no especial, 17-28.

VALENTE, O. (ed.) (1989). Manuais Escolares. Análise de Situação. Lisboa: GEP do Ministério da Educação.

VALENTE, O. (1989). A escola e os valores. (proposé pour publication).

VELOSO, M.G. RODRIGUES, M.J. (1987). A Escola e o Desenvolvimento dos Alunos. Um Ano de Trabalho Com as Escolas. Lisboa: GEP do Ministério da Educação.

1.5
Socialization and Human Rights Education – The Swedish Case

Bengt Thelin

Sweden

SUMMARY

This paper concentrates on social and economic human rights and on the necessity for a global perspective in human rights education. To some degree the paper is a further and more personal development of an "Action programme for internationalisation of the school" presented by the Swedish National Board of Education one year ago, where international solidarity is strongly emphasised.

The wealthy developed countries in Western Europe have special obligations concerning the present global issues of mankind. A more concentrated, transdisciplinary and coherent education is called for. An "education for survival" is very briefly outlined, the basis of which is the *unique* and *absurd* situation of mankind, which must cause a reconsideration of what the most *relevant* content of education today must be. Human rights in the broadest sense is an integrated component of an education for survival together with the topics of peace and war and global ecological issues.

The pure *cognitive* approach is not enough in an education for survival. Both an *affective* and *action oriented* approach is necessary - i.e. depending on the concern and anxiety of the young generation for the future. One "thesis" of my paper is that education lags with respect to the great global issues mankind today is facing.

Finally some examples and experiences from Sweden of an education characterised by a global and solidarity approach is presented.

1.5.2 INTRODUCTION NOTES

According to the letter of invitation from the Directorate of Education, Culture and Sport at the Council of Europe (15 March 1989) to this research colloquy the experts were asked to present paper "on one of the aspects enumerated in the attached information note". The aspects referred to are obviously the five points listed under the heading "4. Aims of the meeting".

Scrutinising these five points one comes - not surprisingly - to the result that they are more or less interwoven and that it is difficult to stick exclusively to only one of them. However, I intend to deal mainly with point 4.4. which reads: "To draw conclusions from research and experiences discussed with regard to future government policy in this area: in particular with regard to: the approach to be adopted by schools (curricular design, educational orientation etc); -INSET (eventually European co-operation in developing joint INSET programmes).

What I am going to deal with is to some degree a further and may be more personal development of an "Action programme for internationalisation of the school" which was presented by the Swedish National Board of Education (NBE) one year ago and for which I had the co-ordinating responsibility. One passage of this programme emphasises the necessity of strengthening human rights education which should be seen as a very important part of education for international solidarity. NBE is the central state authority for the Swedish school system. At the NBE I have a special responsibility for questions related to international and peace education. As for "future government policy" it cannot be taken for granted that the action programme will be endorsed by the government. Anyhow, it indicates ideas within the NBE, the intention of which is to carry the programme out as far as possible.

I also want to stick to the issue *human rights education* and not to a more general discussion of democratic values and how to socialise pupils in this respect. Likewise I will concentrate on the social and economic human rights and especially on the problems these topics entail for education in the rich and industrialised countries. Human rights education is in my opinion an integrated and important part of peace - or still better - of global education.

Referring once again to the letter of invitation, where it is said that "research and experiences ... in your country ... may serve as a basis for your paper", I will underline that my paper will be characterised more by reflections than research, and more by experiences that experiments. The reason for that is the simple fact that we do not have any real research dealing exclusively with human rights education. Some practical development work, however, will be mentioned.

1.5.3 THE SWEDISH CONTEXT

The intention of the following brief overview of the Swedish school system which is still rather centralised is to make the context and the pedagogical preconditions for and expanded human rights education visible.

Both the 9-year compulsory school age (7-16) and the upper secondary school (16-19), to which more than 90 per cent of all students continue either to some of the two or three year courses or to the briefer vocational courses, have centrally issued curricula passed by the parliament [1]. These curricula (Läroplan för grundskolan, Lgr 80, och Läroplan för gymnasieskolan lgy 70) are valid for the whole country. The introductory part of the curricula contains the overall goals and ideological principles of Swedish school education. The predominant place is hereby held by fundamental democratic values. Upbringing for democracy is the basic educational principle.

Beside this and closely connected to it strong emphasis is laid upon such concepts as international understanding and solidarity, peace and human rights. Even here the normative character is clear. Although objectivity and a critical mind are fundamental principles in our instruction the Swedish school is not neutral when it comes to questions of democratic and humanitarian values.

The cognitive elements of these values are defined more in detail in the syllabuses of such subjects as religious knowledge, history and civics.

It is a matter of fact that the Swedish school has had this international character at least since the beginning of the sixties. To a large extent this international interest has been concentrated on third world countries and culture.

Since the end of the 70s and early in the 80s an East-West orientation in internationalisation has become more visible as a complement to the North-South perspective. Related to the advancement of the peace movement, issues like de#Atente and disarmament have become more important even in the schools. The demand for a more consequent and systematic education *on* peace and *for* peace has grown stronger. Later on in my paper I will return to this development. Let me here just mention that the NBE since the beginning of the 80s has been anxious to support and legitimate peace education. One manifestation of this support is a booklet called PEACE EDUCATION-Peace, Liberty Development, Human Rights. It is set out according to the didactic scheme WHY-WHAT- HOW and it states that the aim of peace education in schools "must be to increase pupils' knowledge and awareness of the great survival problems in the world, of war and peace and of man's responsibilities and opportunities for constructing a safer world". (Some copies in English of this material will be available at the colloquy.) I believe that this is the context and the framework for human rights education.

To summarise, I think it is correct to say that the Swedish curricula are more or less permeated with those values and principles, which are fundamental for human rights in the broadest sense.

1.5.4 A GLOBAl ASPECT OF HUMAN RIGHTS EDUCATION

During the last 40 years we have witnessed the coming into being of a very extensive body of norms and standards in the field of human rights. The beginning was made by the Universal Declaration of Human Rights, proclaimed by the UN in 1948. In 1966 this declaration was followed by the UN International Convenant on Economic, Social and Cultural Rights and the Covenant on Civil and Political Rights. Together with and Optional Protocol connected to the last mentioned covenant these three UN documents constitute the International Bill of Human Rights.

At regional level, however, and as early as 1950 the European Convention on Human Rights was adopted under the auspices of the Council of Europe. According to this convention, alleged violations of human rights may be examined by the European Commission of Human Rights and the European Court of Human Rights. It meant that two very effective institutions were set up to supervise civil and political rights.

I can imagine that in all there now exists some hundreds of recommendations, declarations, conventions, etc in the world dealing with human rights from a wide range of perspectives. They represent a rich collection of international instruments aiming to protect mankind from abuses and violations of different kinds [2].

It is an encouraging fact, somewhat of a humanitarian victory, that it has been possible to create such a codex of moral and legal rules. However, legislation is one thing, implementation and information something else. It is in the information work that education and schools have a fundamental role to play. It is worth mentioning in this connection that the UN General Assembly last year decided to launch world public information campaign for human rights. Special emphasis will be laid upon education.

What I would like to concentrate on in my contribution is what role education in the rich Western European countries can and should play, first and foremost concerning the social and economic rights from a global point of view. It does not mean that I think the political and civil rights in these countries can be neglected in education. Not even in my own country, that has lived in peace for 180 years and experienced a very calm non-revolutionary growth of democracy, justice and equity the situation can be looked upon as a "perfect" one. A permanent vigilance towards worrying but still feeble tendencies of racism and political extremism is a necessity. Many of my fellow- countryman have also been surprised at some of the negative judgments for Sweden at the European Human Rights Court.

Nevertheless, I think it is correct to say that the Western European countries have reached such a level of security and stability with regard to human rights that we - at least most of us - could pay more concentrated attention in our education to the global situation and especially to social and economic rights. It i my conviction that we cannot afford to limit education on human rights and what they mean to our own privileged corner of the world. On the contrary, I see it as an obligation for us to realise that rights, which we regard as something natural and self-evident, for a large part of the world population are still unobtainable and utopian ideals or false rhetoric. Dealing with this topic in education confronts us with several serious psychological, pedagogical an moral problems. With all respect for what has been done until now I believe it is true to say that school education has failed to take seriously the global aspects of human rights, ecology and peace. In the next part of my paper I intend to develop very briefly what I mean by an "education for survival".

1.5.5 EDUCATION FOR SURVIVAL

The survival of mankind is nowadays a concept frequently dealt with in science, literature, theatre, film and art. In the history of religions and ideas we can find many examples of thoughts and texts dealing with the final disaster of the world and the end of time. What is *unique* for our generation is that we posses the possibility and the power to let the old myths become reality.

The future of our planet is above all threatened by two facts, the nuclear war and the destruction of the global ecology. A third threat is what often is called the overpopulation of the world. It is a well-known fact that the world population is increasing very rapidly: one billion in 1830, three billions in 1930, five today and eight in 2020. There are, however, reasons for believing that those scientists are right who claim that our planet has resources enough to feed all these human beings. What is wrong is the unjust distribution of the wealth of the world. About twenty per cent of the world population is suffering from hunger and malnutrition. Every minute 30 children die from hunger or hunger-related illness. In the same minute the world spends 1.8 million dollars on armaments. According to the UN report *Our Common Future* ("The Brundtland Report") published by the World Commission on Environment and Development 1987 an action plan for tropical forests would cost *1.3 billion* a year over the course of five years. This annual sum is the equivalent of half a day of military expenditure worldwide [3].

Such facts and figures - the Brundtland Report has many of the same kind - are examples of what I would call the *absurdity* of our time. It is an absurdity which is strongly related to human rights, indicating how these rights - particularly the social and economic ones - are violated in the most appalling way in our world today.

Several studies show that many children and young people feel fear for the future of the world and are sharing a collective pessimism. But not only that: there are also many expressions of disillusionment among young people at the attitude of an older generation who do not care about what is going to happen. Other studies indicate both that questions on nuclear war and environmental threats are regarded as very relevant and important matters by the students and that they feel that school does not deal with these things thoroughly and well enough. The main source of their knowledge of the great global threats seems to be the mass media [4].

Accordingly, a sensible conclusion is that school education has not yet taken up its full responsibility for dealing with this most *relevant* subject - the future of mankind and the threats against it which our generation is facing.

The *unique*, the *absurd* and the *relevant* are the three concepts that constitute what I call an *education for survival*.

I have already said that we in Sweden are lucky enough to have central curricula with general goals and guidelines issued by parliament and binding for all schools education in the country. It is an obvious fact that these goals and guidelines also include normative statements on the necessity of global solidarity and co-operation across both national and ideological frontiers if the inhabitants this planet - the space-ship Earth - are to be capable of solving the great problems concerning poverty, oppression, peace and ecology. Such sentences and statements, decided upon unanimously by parliament, are the formal basis on which an education for survival can be based and legitimated. The important and to some degree controversial thing is how to transform this rhetoric into reality.

As or the legitimacy it is - especially for countries without national goals concerning an international or global education - important to remember that the whole UN ideology is ultimately based on the thought that education is the most indispensable instrument to change the world and "to save succeeding generations from the scourge of war". This fundamental thought is thoroughly developed in the "Recommendation on education for international education" adopted by UNESCO in 1974. On the regional level the Council of Europe plays the same role with its strong emphasis on human rights education, cf. its Recommendation No. R (85) 7 of the Committee of Ministers to member states on teaching and learning about human rights in schools. There also exists a very impressive row of books, booklets, reports etc from seminars and courses published by the Council of Europe on human rights education.

What has happened in Sweden and probably in many other countries as well during the last decade is that an educational movement has emerged characterised by a much stronger emphasis on the great questions of destiny which mankind has to face. The origin and the promoters of this movement are to be found more among individual teachers and students rather than in the educational authorities. In Sweden, however, the above-mentioned Action Programme for Internationalisation of the School, worked out in 1988 by the Swedish National Board of Education, can be looked upon as an official attempt to support this movement. Very briefly and noting both the action programme and school experiences this movement can be described as follows:

In the matters of *content* and *cognitive* components much more attention - as has already been touched upon - has to be given to questions of peace and war, the developing countries, human rights and ecology.

However, there must also be an *affective* component included in any programme of education for survival. Nobody seriously dealing with current world problems mentioned above can remain unaffected when confronted with the absurd priorities and unjust distribution of resources which are predominant in our time and by the fact that we permit this tremendous amount of human suffering to proceed. Speaking about the affective component I have two aspects in mind. *Firstly,* that there is no reason (or right) for education to disguise the reality, although judgment and caution, of course, are highly important, especially in the junior grades. Children and youngsters have the right to know and in fact also to be upset. *Secondly,* feelings of indignation, anxiety and fear have to be channelled into productive tracks and thus used as a power and resource for positive and hopeful activities.

This leads to the third component of an education for survival, namely *action*, which is the natural consequence of empathy and feeling. Children and young persons react vehemently to discrepancies between words and action. What is now encouraging to see is that many schools have started their own action- oriented international projects of different types. I will later on say a little more about how international solidarity can be put into practice as an integrated part of the life in a school.

Nothing of what I have mentioned above as components and characteristics of an education for survival is something entirely new. Nevertheless I think that as a whole it means something new, a new pedagogical dimension and approach. I would like to summarise this in the following points:
1 More weight and scope to be given to the global questions.
2 Not only civics and history but science subjects, languages and practical and artistic subjects can also contribute.
3 As a result, co-operation and joint planning must occur between teachers. Project and thematic studies should be encouraged.
4 In addition to the cognitive element there should also be an affective and normative component.
5 The traditional studies should be interwoven with some forms of practical international and humanitarian action constituting a permanent and legitimate feature of schoolwork.

1.5.6 EXAMPLES AND EXPERIENCES

In this section I would like to present some Swedish examples and experiences relating to what I have said above and also raise some principal questions, which I hope can be of some value for our discussions during this colloquy.

Before that, however, I think a commentary on my terminology in this paper is required. As for the term "education for survival" I am anxious to say that it does not have any official sanction, nor is it used in any official documents in Sweden. It is more or less my own invention. My hope is, however, that my paper can give some evidence of the reason for using it.

I am also aware of the fact that my broad use of the concept "human rights education" can be criticised. Bearing in mind, however, the broad span, which the concept human rights itself has, I think it can be defended. This span manifests itself first of all in the 1948 Universal Declaration and in the two UN covenants in 1966. Let me also remind you of UN declarations like those ones on the social progress and development (1969) the eradication of hunger and malnutrition (1974) and the right of peoples to peace (1984). Another one that must never be forgotten is, of course, the Declaration on the Rights of the Child, which will, as we all hope, soon be "promoted" to a covenant.

That a global aspect of human rights education is not unfamiliar to the Council of Europe can be demonstrated by the European Campaign "North-South Interdependence and Solidarity" and the excellent information kit produced and distributed by the Council in this connection and from which I cite the following passage:

"The intention behind the campaign is to offer the European public a new vision of relations with the third world, with due emphasis on interdependence between North and South.

But the intention is also to ensure that the Council of Europe member states, dedicated to the protection of the individual and respect for human *rights in the broadest sense* (political, social, economic and cultural rights), at all times reserve a place in their policies for Europe's responsibility towards developing countries".

Let me also mention that the Council of Europe's 10th Teachers' Seminar on Human Rights Education had as its theme "human rights education in a global perspective"[5]. Probably other examples as well could be mentioned indicating that a global dimension of human rights education is within the Council's field of interest.

Another term which I have previously used in a broad sense is "peace education". However, if is a well-known fact that peace education is often looked upon as something dangerous and suspect. The truth is, although paradoxical, that if you want to have peace concerning studies on issues like peace and war, disarmament, peace movements, human rights etc you have to avoid the term "peace education". "Global studies" or "global issues in education" might be better alternativaes. Or why not "human rights education" - taken in the broad sense I am arguing for in this paper, provided it includes an ecological dimension as well. The right to clean water and unpolluted air is after all an unalienable human right and an elementary precondition for a life in dignity. Soweto and Seveso are two names that could be mentioned as symbolic for such a broad concept of human rights.

After all, the concept or the term is not the most important but the reality behind, namely that a concentrated and coherent education on the contemporary global issues is carried out. What it all is about are threats against men and their future caused by men themselves.

And now some words about Swedish examples of practical activities in the school related to human rights in "my" broad sense. Rather frequent are charity work and school-twinning with developing countries, co-operation with international humanitarian organisations such as the Red Cross, Save the Children, Amnesty International etc and contacts and correspondence with schools in other countries in the industrial world, preferably those ones with political and economic systems different from one's own. Examples of this kind are correspondence and co-operation between schools in Sweden and Poland on environmental problems, for instance air pollution.

The co-operation with international humanitarian organisations also takes place of a central level. Consequently, there is a permanent working group set up by the NBE with representatives from these organisations. The intentions which lie behind the forming of this group are two: firstly, we want to demonstrate that the Swedish school as a whole is representing and promoting the same humanitarian values as for instance the Red Cross and Save the Children. Secondly, we have some joint publishing activities. One is that we annually publish a booklet enumerating study materials emanating from the organisations. In this way we co-ordinate our efforts of informing the teachers and can avoid duplication. Another is that each year we run a seminar for teacher trainers, who are, of course, a key group for the long term work of implementing international issues in the school.

Important for the action-oriented work has been the fact that the NBE during the last few years has been able to give some financial support to schools with particular interest in these issues. They are small sums only, from 5,000 to 25,000 Swedish crowns (i.e. about 500 to 2,500), but obviously this money has an encouraging and stimulating effect and a legitimating one as well. Devoted teachers in this area often feel alone. Not rarely they are - with some ironic touch - called "fiery spirits". Official and financial support gives them a better status. Of special value is that the Olof Palme's Memorial Foundation allocates an annual sum to be distributed by the NBE to schools especially committed to initiatives on international contacts, international understanding and antiracism.

A special example of action is the spontaneous mobilisation and action by students in aid of refugee school - fellows threatened by expulsion from the country. Of course, here we are facing a delicate and controversial problem relating to human rights calling for a great portion of judgment from school and immigrant authorities. However, not too rarely the "interventions" of students - sometimes supported by their teachers - have resulted in a positive and altered decision by the National Immigration and Naturalisation Board or by the government.

It is our hope and intention that schools dealing with international solidarity projects will give a valuable contribution to the development of teaching strategies and methods and to some extent play a pilot role.

Consequently they have to deliver reports on their projects (6). For simple reasons a more qualified evaluation cannot be expected. It would call for much more resources than are available at present. Nevertheless, the ongoing "research" means the collecting of experiences, which in the long run might be of interest for the research and development work. In any case it seems to be clear that activities of this kind are good albeit unconventional learning experiences, which the sometimes reluctant school authorities on different levels in the future hopefully will sanction and approve as legitimate ones. There are also indications that activities like the above described create a positive atmosphere in a school and a feeling of solidarity and good fellowship.

Of course, there are several obstacles to overcome and many questions to answer before we can with any justification talk about something like an "education for survival". In conclusion I would like to mention and very briefly comment on some of them.

1 A classical question/objection is related to the overloaded curriculum. For me the answer is given: if you really want to - you can. That it must entail some reconsideration of the traditional subject-matter goes without saying. What it finally depends on is which knowledge and which skills we think are the most central ones for mankind - now and in the future. Thinking in new ways often is a painful process.

2 Issues of human rights, peace, ecology etc, can be very controversial, still more if we in the school not only talk but also have the intention to do something. Is there not a risk of "indoctrination" and "politicising"" of school and education? I think there *is* a risk. That is why these questions must be handled carefully and with sound judgment. On the other hand, if there exists - which as mentioned is the case in Sweden - a curriculum that emphasises humanitarian values, international solidarity etc the head and the teachers have legitimation. These values are in our time too important to be dealt with as pure rhetoric.

3 A third question is of a more practical kind: how to get the teachers to co-operate across the subject boundaries and to apply cross-curricular teaching? I, for my part, cannot find any other answer than that we have to be patient and rely on the long-term effect of the initial and further training of teachers. But we have to begin with the teacher trainers themselves! Could it be possible that the Council of Europe takes some special action for that group - and for the head teachers as well, which are the second key group for bringing about innovations?

Let me as my strong personal conviction say that those of us belonging to the wealthy countries in Europe have extraordinary obligations as for human rights education in a global perspective and for an education both *on* and *for* human rights. We have to teach our students and ourselves to be something more than mere spectators, if we want to be regarded as trustworthy. In 1989 the trinity Liberty, Equality and Fraternity is of current interest. In the history of human rights they have been of fundamental importance. I think there are reasons for saying that we in our countries today have reached good results in liberty and equality. As for fraternity taken in a global sense we still have a very long way to go.

Finally, permit me to formulate as a kind of very personal *Pia Desideria* that human rights in its global sense in our countries could share at least one per cent of the interest, debates and resources at present devoted to the European Common Market.

NOTES

1 Adult education is not dealt with in this context, although it has to be

said that international studies and peace and human activities are of great importance in many folk high schools.
Enclosed is a brief draft of the Swedish school system. About 98% of all Swedish youngsters are students in the official school system.
2 Of "Human rights. A compilation of International instruments", UN, New York, 1988.
3 Our common future, p. 297.
4 Cf Bergström, 1984, Billesböle 1986, Chivan 1986, Randalen 1986. Worth mentioning in this connection are also two inquiries among teachers who are also parents on their opinion on the content of the Swedish comprehensive (the 9-years compulsory) school. One of several interesting results was the high ranking of Peace and International Understanding, cf Marklund (a) and (b).
5 Cf report from Are, Sweden 9-14 August 1987.
6 One report is enclosed as Annex 2.

1.5.7 BIBLIOGRAPHY

Bergström A, Barn och kärnvåpen (Children and nuclear weapons) (I Värld och Vardag) Tikskrift för oä, no, so. Nr 3 1984 Liber Utbildningsförlag.

Billesbölle P m fl. Börns angst for krig med kernevåben (Children's fear of nuclear war) (Leageforeningens forlag, Köpenhamn 1986)

Bjerstedt A, Peace education in different countries (Educational Information and Debate, 81) Malmö: School of Education, 1988, Sweden

Bjerstedt A, Peace education news - from Sweden and elsewhere. Reprints and Miniprints (Malmö, Sweden: School Education), No. 620, 1988.

Chivan E, Youngsters and nuclear weapons. A research by Nuclear Psychology Program, Harvard Medical School, Cambridge, Massachusetts, USA, 1986.

Council of Europe, European Teachers' Seminar on "Human rights education in a global perspective". Are, Sweden, 9-14 August 1987. Report by Sanchia Pearse, Strasbourg 1988.

Council of Europe, Human rights education in schools: concepts, attitudes and skills by Derek Heater, Strasbourg 1984.

Council of Europe, Recommendation No. R (85) 7 of the Committee of Ministers to member states on teaching and learning about human rights in schools, Strasbourg 1985.

Council of Europe, the European Convention on Human Rights, Strasbourg.

Marklund I, (a) Allmähetens bild av grundskolan och dess innehåll. En SIFO-undersökning oktober-november 1987, (a public opinion poll on the Swedish comprehensive school and its contents), Swedish Board of Education, Report 88:2.

Marklund I, (b) Att vara lärare i grundskolan. On hur lärare ser sin yrkesroll och arbetsmiljö, (to be a teacher in the comprehensive school) Swedish National Board of Education, Report 88:37.

Our Common Future ("The Brundtland Report") pusblished by the World Commission on Environment and Development 1987, Geneva 1987.

Raundalen T S och Raundalen M, Barn i Atomåldern) (Children in the nuclear age). Vad Säger vi till barnen och vad säger barnen själva, Prisma, Stockholm 1986.

(Thelin B), Action Programme for Internationalisaton of the School. Swedish National Board of Education 1988.

Thelin B, Peace education. Peace, Liberty, Development, Human Rights. Swedish National Board of Education, 1986.

UNESCO, Recommendation concerning education for international understanding, cooperation and peace and education relating to human rights and fundamental freedoms, Paris 1974.

UN Universal Declaration of Human Rights, New York 1948.

UN International Covenant on Civil and Political Rights, New York 1966.

UN Optional Protocol to the International Covenant on Civil and Political Rights, New York 1966.

UN Universal Declaration on the Eradication of Hunger and Malnutrition, New York 1974.

UN Declaration on the Right of Peoples to Peace, New York 1984.

UN Declaration on the Right to Development, New York 1986.

UN Human Rights. A compilation of international instruments, New York 1988.

1.5.8 APPENDIX 1: THE SWEDISH SCHOOL SYSTEM
(as described by the Swedish National Board of Education)

All children in Sweden start school the year they are seven. Before this they may have attended preschool, which is voluntary. Compulsory schooling lasts for nine years, after which 90 per cent of school-leavers go on to upper secondary school, from which about 25 per cent proceed to some form of post- secondary education. Several million Swedes over the age of 18 also take part in various forms of adult education.

Preschool education

All children aged 6 and over are eligible for pre-school education, which is optional for the children but compulsory for municipal authorities. Responsibility for preschool education is vested in the National Board of Health and Welfare.

Compulsory school

Roughly one million pupils, i.e. all children between the ages of 7 and 16, attend the nine-year compulsory school, which is divided into three levels: junior, interme-diate and senior. The great majority of these schools are run by municipal authorities and are free of charge. Nor is any charge made for teaching materials, school meals, health care or school transport (for children living a long way from school). There are also a very small number of private schools.

Upper secondary school

Upper secondary school is divided into 25 lines of two, three or four years' duration. Some of these are vocational, while others lead on to further education. Upper secondary school also includes more than 500 directly vocational specialised courses of varying duration. About 25 per cent of upper secondary school-leavers go on to higher education. The remainder enter employment, possibly entering higher education or some form of adult education later on in life.

Adult education

There are many different forms of adult education in Sweden, viz municipal adult education (conferring essentially the same qualifications as compulsory school and upper secondary school), folk high school, labour market training or some form of study circle activity. Adult education is often free of charge.

91

1.5.9 APPENDIX 2: REPORT ON THE "HELP MOZAMBIQUE"

PROJECT (funded by the Olof Palme Memorial Fund)
Report on "Help Mozambique" project which received funds from the
"Olof Palme's Memorial Fund"

County: Skaraborgs County

School: Alléskolan, Vara

Level: Senior level of compulsory school

Number of pupils/classes taking part in project: 400 pupils, 15 classes

Name of project: Help Mozambique

Project leader: Wolmar Bengtsson, Lennart Bornäs, Said-Ove Rydén

Grant allowance from "Olof Palme's Memorial Fund": Skr 6,000

Allowances from other sources: Skr 0

Use of funds:		
	Photocopies	64 Sw. cr
	Hire of videogramme	400 Sw. cr
	Advertising	2,677 Sw. cr
	Packing	1,150 Sw. cr
	Specialist help (Hans Abrahammsson, Susanne Fjällemark)	1,500 Sw. cr
	Total	5,971 Sw. cr

Aim of project: We wanted to fulfill the goal of the curriculum in a concrete
way with regard to teaching in international questions. We look upon the
project as a possibility for the pupils and staff of the schools to work in a
practical way to maintain a readiness to help when help is needed. Also,
through teaching about South Africa, give an example of one of the world's
problem areas.

Project description: The school year 88/89, spring term. To support the
continual teaching in international questions, we have worked with South
Africa, specifically Mozambique, in the following way:

- Different models of lessons have been offered Allëskolans teachers.
- Teaching material has been offered the junior and intermediate levels of compulsory school in Vara.
- Pupils from the senior level of compulsory school have held lessons for their younger junior and intermediate school companions - about 20 classes.
- Theme days with specialist help.
- Work for Mozambique:

    * daywork for pupils
    * collection of clothes
    * collection of money
    (see report enclosed with this year's application form).

Results: The work has led to a deeper understanding among the students and staff about the situation in South Africa. Involvement with people in other lands has grown. Even the general public has received greater knowledge and understanding, through newspaper articles and "mouth to ear" methods:

- 30 m3 schoolmaterial to the secondary school in Namuno.
- 60 m3 clothes and shoes to the district of Namuno.
- Skr 27,000 to use for the buying of school material for the secondary school during next year.
- Skr 17,000 from Arentrops junior and senoir level schools who had their own daywork project to raise funds.
- A support association for Mozambique - and for our work - has been formed in the municipality.

Effects of propagation: The work has led to an increased sureness among many pupils. They have had the possibility to take inititiative and show that they can shoulder personal responsibility. Co-operation between students and staff has grown stronger.

Final comments: We are counting on continuing the work which has been started in a like manner, and hope that even in the future to receive means from the "Olof Palme's Memorial Fund" to cover the costs of this activity. We wish also to continue sending everything we have collected, without any reduction to aid our sister-school and the people in Namuno.

Ernst Lundquist
Headmaster
Alléskolan, Vara

1.6
Human Rights and Moral-Cognitive Development

Heinz Schirp

Federal Republic of Germany

1.6.1 INTRODUCTION

In recent years there has been considerable debate in the Federal Republic of Germany on the assumptions, aims and concepts of value-education.

Put briefly, this controversy proceeds from two premises. On the one hand there is a demand for traditional virtues with their emphasis on socially desirable behaviour and norms to be transmitted. On the other hand schools and educators are expected through their stress on the notions of emancipations, involvement and political participation, to enable pupils to be critical of society.

It seems to us that neither "indorctrination and the learning of virtues" nor "conflict-theory and a critical distancing of the society" not to speak of a "relavism of values" is the right direction for schools and education to take. What follows, therefore, is the description of the development of a moral-cognitive sense as a genuine alternative to these.

At present in North Rhine Westphalia we are engaged in a research project

"Human Rights Education and Moral - Cognitive Development"
Four caricatures - four critical aspects - four questions - four answers

1. "The forgotten pupil"	2. "The different perspectives"
In most approaches to moral education and values education the "standpoint" of the students is forgotten and is not considered. • Where is the developmental and cognitive basis to understand values and to teach them?	Often the moral perspective of the teacher is not that of the student. The one is "principle-orientated" the other "group-orientated". • What can we do for teachers to understand the moral-cognitive perspective of their students?

3. "Closeness to life"	4. "The measured student"
Often the interests of teachers and their ways to teach or indoctrinate values and virtues don't correspond to the interests of the students. • Where are the relations between values education and the interests of our students?	Some teachers are afraid that the theory of moral-cognitive stages leeds to moral measurement. • function can moral-cognitive stages have in teaching and educating processes?

with a few schools which is based upon the developmental-psychological status of the children and adolescents involved. We are working closely with the Pedagogical Institute of Fribourg University in Switzerland to develop and test materials which cultivate a sense of moral judgement. In addition to this, we are experimenting with democratic structures which, through the involvement of the pupils, should constitute an integral part of school life.

This report can do no more than sketch the assumptions on which this work is based and give an indication of the initial findings.

1.6.2 BASIC ASSUMPTIONS

The following is an outline of five elements on which our concept of human rights- and values-education is based.

1.6.2.1 Society

It is essential for a democratically constituted community to be based a consensus of norms and values. Can we say that such a consensus exists, or are State and Society not in the oft-cited crisis of legitimacy, in particular in the field of human rights? In the face of mounting economic, ecological and social crises, it is surely the case that solutions to conflict are sought more often in a short-term technological pragmatism and economic advantage than in a consideration of rihts, interests and needs. The question of an appropriate future produces increasingly contradictory responses. The present uncertainty as to the correctness of norms can, for instance, be traced to the fact that the general system of values no longer accords with people practical interests. [1]

1.6.2.2 Young people

Adolescent-sociological analyses show that the uncertainty over norms which is communicated through the society leads tovery diverse political behaviour. A large proportion of young people refuse to become involved in the democratic processes of the society, feeling the nature of the system itself precludes their making a valid contribution to it. This leads to a retreat into their private sphere, into subcultures or into 'scenes' based on consumption which might be summed up in the words of a current hit-song 'Don't worry, be happy'. Side by side with this, however, exists the potential for protest and demonstration which at the moment makes common cause with radical-democratic challenges to state policies ('the arms-rac', 'the environment', 'apartheid', 'opposition to the carrying-out of a census'). [2]

1.6.2.3 Ethics

One of the principal questions that permeates the philosophical discussion on values and moral education is this: Given a realistic value-ethic, is it possible to construct a catalogue or even perhaps a hierarchy of values? If that were to be the case, then education on moral conduct could look to this scale of values for guidance. The rule could then be that in the case of a conflict of norms and values, one should orientate oneself on the values on the next highest step of the scale. All attempts at establishing such a scale of values which operates independently of a context have proved futile. Futile above all because such a hierarchy cannot be established in a pluralistic, ideologically and religiously neutral, democratically constituted society. Furthermore, the strict adherence to such gradations of values would not lead to that which for a democracy is the 'condition sine qua non' - the individual and common commitment to good judgement and freedom of action. Democracy depends largely on this judgement.

Modern discourse-ethics has returned to the formal principle of examining values. In Kant's philosophical tradition this principle aims to make the individual examine the motives underlying his conduct to ensure they are compatible with the dignity and rights of <u>all</u> other people. The individual must reconcile his conduct with the interests of others if he is to ensure its moral legitimacy.

1.6.2.4 Developmental psychology

Before embarking on an attempt to communicate to children the mening of human rights as a universal expression of values, it must be asked whther this is possible and, if so, how. Can children and adolescents comprehend such a structure of values? The developmental-psychological work of Jean Piaget and Lawrence Kohlberg demonstrate how important it is to take into account the moral-cognitive stage of development children and adolescents have reached.
Piaget divides the development of moral judgement into three stages:
- the observance of given rules; (moral realism)
- the acceptance of agreement in the peer group; (cooperative sense of justice)
- the training of the independence will; (consciousness of the concept of justice). [3]

The work of the American psychologist and teacher Lawrence Kohlberg, which is rooted in these ideas, will be dealt with in more detail later, as it forms the basis for our practical studies in the school.

1.6.2.5 Teaching and schools

Empirical research on the question "What is a 'good' school?" has shown an irrefutable relationship between educational concepts and the quality of the school. The characteristics of a good school are that

- it connects the learning of individual subjects with general educational conduct,
- there is a high degree of consensus among staff-members on the educational philosophy of the institution
- it sees itself as a material constituent of the lives of its pupils, and thus imparts a sense of relevance through the social make=up of the school environment,
- a benevolent social atmosphere prevails.

In schools which lack such a pedagogical consensus, teacher dissatisfaction nd pupil indifference can often be observed. Brookover et al. speak of 'a sense of futility'. This, in turn, affects the pupils' performance. [4]
The educational structure and learning performance of a school are dependent on one another. This presupposes that education is an integral part of school-life and classes. Studies on 'the hidden curriculum' have shown that all the admirable norms which we wish to impart to our pupils in diferent variations and in different subjects remain ineffective if they are at variance with the experience of everyday life in the school and in the classroom. In summary, the aims or four study on values-education are:
1 It should enable the pupils to engage in a retional discussin on values and lead to an analysis of the justification of values.
2 Thereby, it should contribute to the development of a sense of moral-democratic judgement.
3 It should lead to the questioning of the validity of norms and values for all individuals.
4 It should take into account the capacity to reason and judge which will depend on the pupils' stage of development.
5 It should help to improve democratic communication and participation in the school as a whole.

1.6.3 THE DEVELOPMENT OF A MORAL-DEMOCRATIC CAPACITY

1.6.3.1 The stages of moral development

The school-based research project in North Rhine-Westphalia is based on the concept of moral-cognitive development which springs from the work

THE SIX STAGES OF MORAL JUDGMENT

Content of Stage			
Level and Stage	What is Right	Reasons for Doing Right	Social Perspective of Stage
Level 1: Preconventional. Stage 1. Heteronomous morality	Sticking to rules backed by punishment, obedience for its own sake, avoiding physical damage to persons and property	Avoidance of punishment, superior power of authorities.	*Egocentric point of view* Doesn't consider the interests of others or recognize that they differ from the actor's, doesn't relate two points of view. Actions considered physically rather than in terms of psychological interests of others. Confusion of authority's perspectivews with one's own.
Stage 2 Individualism, instrumental purpose, and Exchange	Following rules only when in one's immediate interest, acting to meet one's own interests and needs and letting others do the same. Rights is also what is fair or what is an equal exchange, deal, agreement	To serve one's own needs or interests in a world where one has to recognize that other people also have interests.	*Concrete individualistic prspective* Aware that everybody has interests to pursue and that these can conflict: right is relative (in the concrete individualistic sense)
Level II: Conventional Stage 3. Mutual interpersonal expectations. Relationships and interpersonal conformity	Living up to what is expected by people close to you or what people generally expect of a good son, brother, friend, etc "Being good" is important and means having good motives, showing concern for others. It also means keeping mutual relationships such as trust, loyalty, respect, and gratitude	The need to be good person in your own eyes and those of others caring for others, belief in the Golden Rule desire to maintain rules and authority that support sterotypical good behavior.	*Perspective of the individual in relationships with other individuals* Aware of shared feelings, agreements and expectations which take primacy over individual interests. Relates points of view through the concrete Golden Rule, putting oneself in the other guy's shoes. Does not yet consider generalized system perspective.
Level II: Conventional Stage 4 Social system and conscience	Fulfiling duties to which you have agreed; laws to be upheld except in extreme cases where they conflict with other fixed social duties. Right is also contributing to the society, group or institution.	To keep the institution going as a whole and avoid a breakdown in the system "if everyone did it", imperative of conscience to meet one's defined obligation (Easily confused with stage 3 belief in rules and authority)	*Differentiates societal point of view from interpersonal agreement of motives* Takes the point of view of the system that defines roles and rules considers individual relations in terms of place in the system.
Level III: Postconventional, or Principled Stage 5 Social contract or utility and individual rights	Being aware that people hold a variety of values and opinions and that most of their values and rules are relative to their group. Relative rules usually upheld in the interest of impartiality and because they are the social contract. Some nonrelative values and rights (e.g. life and liberty) must be upheld in any society and regardless of majority opinion.	A sense of obligation to law because of one's social contract to make and abide by laws for the welfare of all and for the protection of all people's rights. A feeling of contractual commitment, freely entered upon, to family, friendship, trust, and work obligations. Concern that laws and duties be based on rational calculation of overall utility, "the greatest good for the greatest number."	*Prior-to-society perspective* Rational individual aware of values and rights prior to social attachments and contracts integrates perspectives by formal mechanisms of agreement, contract, objective impartiality and due process. Considers moral and legal points of view, recognizes that they sometimes conflict and finds it difficult to integrate them.
Level III Postconventional, or Principled Stage 6 Universal ethical principles	Following self-chosen ethical principles. Particular laws or social agreement usually valid because they rest on such principles, when laws violate these principles, one acts in accordance with principle. Principles are universal principles of justice equality of human rights and respect for the dignity of human beings as individuals.	The belief as a rational person in the validity of universal moral principles and a sense of personal commitment to them.	*Perspective of a moral point of view from which social arrangements derive* Perspective is that of a rational individual recognizing the nature of morality or the fact that persons are ends in themselves and must be treated as such.

of the American psychologist and teacher Lawrence Kohlberg. We wish to sketch these ideas, as they ar esential for an understanding of our work and its findings.

The empirical research of Kohlberg and his lollowers can be summarized as follows:

- Moral-cognitive development proceeds in stages. Each stage is defined in terms of the moral reasoning which it employs.
- Stages form an invariant sequence. Individual stages cannot be skipped.
- The stages form a hierarchy with each higher stage reintegrating and 'elevating' the lower ones.
- The stages are universal, i.e. they retain their validity in the most diverse cultural settings.
- After a stage 0, the pre-moral stage (roughly corresponding to the infant which acts and reacts purely on the basis of wants), we can distinguish three levels of moral development.
  -the preconventional level
  -the conventional level and
  -the post conventional level.

Each level can be further divided into two stages so that altogether there are six stages. These can be seen in the table. [5]

1.6.3.2 Stimulating the capacity for judgement

From this anslysis of the stages we cn devise a framework for pedagogical action in the school.

- Firstly, techers must recognize the level of development the pupils have already reached; they must be aware of the reasoning-structure, the type of argumentation which pupils will apply to problems and conflict.
- Secondly, they must confront the pupils with moral conflicts and value-dilemmas.
- Finally, they must bring their pupils to tackle argumentation and analysis in forms which lie above their present levels of reasoning.

The pupils can also be confronted with stories, situations and experiences in which people have te decide between two competing values - e.g. between life and property; friendship and law; self-interest and solidarity; law and human-rights.
In each of these cases, the pupils indicate how the person should behave. It is important that the pupils be able to explain *why* the people should react

in one way rather than another. It is also part of the teacher's task to confront the pupils with reasons from the next highest stage so that the cogency of the arguments can be questioned.

This stimulation of the ability to make moral judgements through being confronted with argumetns from the next highest stage forms the core of our project.

Behind this lies the acknowledgement, that each stage must be seen as a structured whole which springs from the subjects themselves. It is, therefore, impossible to 'teach' a higher stage - it must emrge as a rsult of the interaction between the subject and social conflict. In fact empirical studies show that children do not retain this 'learned' reasoning which is not internalized, when questioned, they quickly return to their 'own' level of moral judgement.
A central difficulty which arises here is the simple fact that the ability to engage in a moral discussion does not automatically lead to moral behaviour - insight and conduct are two different things.

A second, related, criticism points to the fundamental distinction between psychological research and practical pedagogical considerations and speaks of a 'psychologist's fallacy'. Put bluntly, not everything the psychologist finds important will be significant for the teacher.

In practical terms this means: Moral judgement can and should not be learnt in the school from fictional texts; it must be based on problems and conflicts arising from the pupils' area of experience. Ultimately the school's objective should not merely be to deal with abstract ideas; it should be more interested in awareness, perception, skills and knowledge as the basis for decision-making and problem-solving. Thus moral-cognitive dilemmas and problems should be interwoven with school subjects and the content of lessons.

1.6.3.3 Just community

The rationale behind the development of a sense of moral judgement must ultimately be that it influences conduct. Therefore the school environment should be a source of experience inwhich pupils can learn to take part in the decison-making process.

Durkheim, Piaget and Dewey have all drawn attention to the fact that a decisive reason - if not *the* decisive reason - for children's moral criteria shifting their focus from the external moral world of adults to cummunally deveoloped and accepted standards, is the evolution of rules in the course of living and working with peers. The 'hidden curriculum' also comes into play here. If pupils who can discuss theoretical moral dilemmas in lessons and devise solutions which are considered fair by all concerned, then recognize that such solutions are disregarded in school life, indeed that school life appears to be organized on totally different principles, they could conclude that reasoned moral judgement is just so much idealistic talk, divorced from reality.

Kohlberg brought this to its logical conclusion. The establishment of a just-community school in 1974 represents an attempt to give pupils a real democratic say in the running of the school. This should ensure tht the connection between the development of a sense of moral judgement and social conduct will be made. This form of just-community school is known as a 'cluster school', which means that pupils from different age-groups and courses in the larger school take part in this programme on a voluntary basis. Thus this form of organization is a 'school within a school'. It underlying concepts are:
1 The school should be run democratically, with staff and pupils each having one vote in the making of school decisions. The consequence are that: (a) rules are arrived at on the basis of agreement between teachers and pupils rather than being handed down from above; (b) all the important questions of administration, discipline and school policy are decided in a weekly face-to-face full-community meeting.
2 Both the administration and the community meetings should use fairness and moral criteria as a basis for reaching decisions.
3 The school curriculum, especially in the area of social studies, contains discussion designed to stimulate moral development and lays particular emphasis on an understanding of the ideas of democracy, rights and justice. The discussions in lessons of morals, rights and democracy are connected to the decisions reached in the community meetings as well as to the relation of the school to the entire school-system and to the society at large." [6]

With these three elements
- development-oriented series of stages
- moral intervention as a means of stimulating moral judgement
- a just-community school as an attempt to combine the structures of judgement and behaviour

I have tried to sketch the approach of our project to moral- democratic education. What follows is a description of the project in practice in the school and the initial results we have observed.

1.6.4 THE PROJECT IN PRACTICE

The practical and research work of the DES project centres on three schools - a 'Hauptschule', a 'Realschule' and a 'Gymnasium'. (These are two forms of middle school and a grammar school). The approach already sketched is being tried out here. In a second phase other interested schools- of which there are a considerable number - can be supplied with appropriate suggestions and help in the form of materials, handouts and advice. After a thorough introduction to the concept of moral-cognitive development, the school taking part have each held planning conference where the scope and aims of the activities have been defined.
I intend to highlight the features which have proved to be central for all schools and which should therefore be transferable. (7)

1.6.4.1 The dilemma-approach

It was decided in all three trial schools to examine the content and aims of the subject curricula with an eye to values-conflict. The following issues emerged:
rules, laws, conscience, feelings, emotions, authority, rights, agreement, trust, reciprocity, punishment, justice, life, property, truth, sexuality, eroticism, solidarity, utility, profit.
It became clear in the discussions that the subject curricula offered a rich source of values-conflict which wuld be relevant both to the general education and the specific subject needs of the learners.

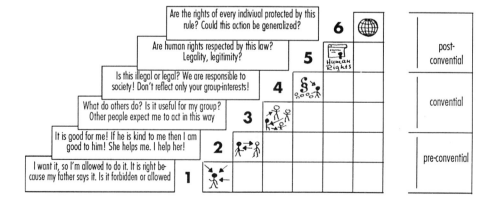

In the individual planning groups it was decided to systematically intro-
duce dilemma-discussions in the following subjects:
- German (literature classes)
- History (development and change of values-systems)
- Politics/Social Studies (values and vested interests)
- Biology (the environment, health, man and nature)
- Geography (responsibility for the environment)

In the participating schools, groups based both on subject areas and on the
wider curriculum were and are analylsing schoolbooks, pupils' workbooks
and current materials from a values-conflict standpoint; -adapting suitable
materials for use in lessons;
identifying the framework of moral levels inherent in the material as an aid
to dilemmag-discusion;
developing methodical aids, impulses and suggestions.

Many lessons have been recorded in sound and on video to help in their
subsequent assessment. Simple aids to the documentation of the topics,
aims and experience of these lessons have been developed. The practice of
dilemma discussions has been considered in subject meetings and
workshops and these have, in turn, produced further suggestions. These
'dilemma' lessons are so distributed that two lessons of this kind can take
place in different subjects in each week.

1.6.4.2 The school community

The idea of the Just-Community-Cluster-School as conceived by Kohlberg
and realized in certain American schools, had to be modified to take into
account the organizational and legal circumstances of schools here. The
basic intentions are comparable, however, Community meetings will take
place in all the trials schools,
- in which, for instance, all classes of a year or a two-year group take part,
- the topics have already been prepared by the pupils in their classes,
- the problems and conflicts arising from school life which the pupils wish
  to see discussed, can be dealt with,
- they can be lead and largely independently organized by the pupils
  themselves,
- pupils and teachers are equal partners in the discussion,
- the pupils can discuss communal school-problems and seek effective
  rules for conflicts
rules can be sought on the basis of their soundness and validity for all,
– the effects and results of self-formulated rules of behaviour can be
  critically observed.

These community meetings will also be recorded and documented, insofar as this is possible.

1.6.4.3 Preliminary experience and results

At present we re still in the experimental phase in the schools, thus we are not in a position to make definitive statements on the basis of the empiricial research conducted so far (questionnaires on the moral-cognitive structure and the school-atmosphere, interviews with pupils). Therefore I will restrict myself to an evaluation of our own experience and a description of the results apparent to a participant-observer.

- The concept of moral-cognitive development has a mobilizing and sensitizing effect on teachers.

We know from empirical studies on school-quality that the teacher is a decisive factor when the question of values and norms is raised. The concept of moral-cognitive development is of interest to teachers not least because it gives them a pedagogical instrument to view judgement-formation from the perspective of the pupils and to adjust their teaching accordingly.
In this way teachers come to a better understanding of their pupils. Almost all teachers have picked up this sensitizing aspect as particularly important. This concept leads simultaneously to a greater emphasis on teh school's wider educational role; it opens up a discussion on the aims and methods of education and reinforces pedagogical commitment. It is precisely this debate on values-education amongst staff-members that appears to be particularly effective in promoting the development of schools.

- The concept promotes the view that for effective learning, the link between social and academic skills must be forged.

The realization that questions of values are an essential part of the individual subjects challenges the subject teacher to confront the pupils with such problems and to help them refine their abilityto judge.
There is good evidence to suggest that the role of the teacher also undergoes a broadening from that of 'subject teacher' to 'educator'. Pupils, made aware that solutions have a human dimension, realize that expertise and values are inseparable elements of a decision.

- Education can become an integral part of lessons and school.

The opportunity of democratic participation has significance for both teacher and pupil. For the teacher, it involves seriously the interests, conflicts and needs pupils encounter in the school and assisting in their coming to terms with them. Here we observe another change in the teachers' role: they must justify decisions and set them against pupils' legitimate suggestions. Such a discussion on equal terms is a part of a democratic culture within the school. For the pupil, it means that their ideas and concerns are taken seriously. They learn to take the interests of others into account when considering solutions to conflicts and demanding changes; they learn to question the acceptability of a regulation to fellow-pupils. This empathizing, seeing the problem through the eyes of another, is a prerequisite of a democratic sense of judgement. At the same time an attempt is made to narrow the gap between a coginitive insight and making the appropriate adjustment to personal conduct and behaviour.

- The limits of and structure resistance to participation in the school are revealed.

It should not be disguised that there are problems associated with the moral-cognitive approach.

Pupils find that the possibilities of putting their ideas into practice are limited. In many cases there are legal obligations binding the school administration, teachers and pupils. This can lead to resignation summed up in the reaction: "We can't change what's really important anyway!" There is a real danger that participation in teh school lapses into nothing more than a charade.

There is the danger that dilemma-discussions can become rituals if the method is applied in a sterotyped way.

Moral discussions and the formal involvement of pupils in community meetings are limited in their effect. It is more important that the schools themselves offer alternatives to the usual school and classroom experiences. They should innovate pupil-oriented forms which motivate pupils to appreciate new possibilities of learning and conduct.

Personally, I feel that it will be necesary to link the moral- cognitive developmental approach to a broader pedagogical one. This second approach must be characterized by an opening of the school in the direction of Community Education.

On one side	On the other side
• Moral-cognitive dilemmas in subject lessons	• Pupil-oriented learning, projectwork
• Encouragement of involvement in shaping lessons	• Learning 'on the spot', in the community, opening of the school through working with persons, groups, experts from the area
• School-community meetings and participation in all areas of school life	• The school as meetingplace

Both approaches support one another and help to put subject-learning, values-education and social concern in a broader context.

1.6.5 THE MORAL-COGNITIVE APPROACH AND HUMAN RIGHTS

In Kohlberg's delineation of levels, individual and human rights belong to the postconventional level. (See Table, §1.6.3.1).
The postconventional level describes the fundamental values and rights governing the regulation and laws valid in a state. 'Life', 'liberty', 'equality before the law', 'human rights' are examples of values against which the validity of a society's legal foundation could be measured.

Levels 5 and 6 undoubtedly demand extremely high levels of argumentation and reflection. The question has therefore often been pesed as to whether schools can possibly bring pupils to a postconventional level.
In our experience, most pupils in the lower secondary school can be placed somewhere on the moral-cognitive levels 2 and 3. How, then, can a framework of human-rights education be constructed which connects the (pre)-conventional and the ambitious postconventional standard? The following elements can be identified from our approach.

The starting-point for such a development-oriented approach is the moral-cognitive stage of the pupils and the
- experience
- influences
- social group
on which they have based their previous values-decisions. The values-decisions of levels 2 and 3 are principally based on one's own interests and the assumption that other members of the group will also act in their own legitimate interests. Correct conduct is therefore seen as a fair balance of interests. At the same time the interests of the other members of one's own social group must be considered.

Level 4 introduces a 'system perspective' to some extent. Correct decisions and conduct spring from the belief that there are rules and laws valid for the entire society, not just for one's own social group.

Level 5 goes further. Here the assumption is that 'legality' (laws, state regulations) and 'legitimacy' (human rights, justice) are associated with one another.

A revealilng connection between levels can be made if values-conflict (dilemmas) and their solutions can be considered to see
- if they can be applied to other social groups, communities or state regulations (transferability)
- if they take into account the interests and needs of all other people (generalization).

Given this need for ratinalization, teaching methods are particularly important. Moral-cognitive informed teaching must help the pupil
- to make a ruling based on the specific situation of individuals of other social and cultural communities, to acquire knowledge about their situation and needs in order to be capable of making a reasonable judgement (change of perspective)
- to put him or herself in the position of those individuals affected by such a ruling, to understand and take seriously their difficulties and needs (empathy).

The consideration of these three stages (assumptions based on specific levels, development aims, development methods) with a view to teaching, can be presented in the following way.

Based on these didactic reflections we have tried to develop - together with the teachers of the above mentioned schools - methodological aids and materials for the various lessons. The aim of these materials is to highlight the relation.
- between the moral perspective and judgement of the students on the one side and those of other individuals and social groups on the other
- between values and value decisions which are acceptable ourselves and those which are acceptable to people of different cultures and to humanity in general.

In these materials the values conflicts are again presented as moral dilemmas which have to be solved.

Moral awareness and the reasoning-structure of pupils are based

on their own EXPERIENCE	on their own SOCIAL GROUP	on their own INFLUENCES
In many aspects of life regulations play an important part.	Norms and rules determine the life in my group, too.	Regulations also affect me directly.
In different questions of values. there are different rationalizations for moral decisions.	it is best when a regulation takes the interests of all members of the group into account.	They influence my conduct towards and my relationship with the other group members.
I must establish the cogency of the reasoning in each case.	I must take part the development of such regulations.	I must reconcile my interests to those of others.

CHANGE OF	FORMATION OF

Subject knowledge and insights into the specific living conditions of individuals of other social and cultural communities.

PERSPECTIVE	EMPATHY

TRANSFERABILITY	GENERALISATION
Not all regulations can be tranferred to other individuals, groups or communities	Each regulation cannot be borne equally well by all those affected.
Do values decisions take the cultural, political and ideological identities of others into account?	Values decisions should be examined to see if they take the needs of all those affected into account.
What must I, or must we put up with as a result of such a values decision?	Can I wish this regulations should apply to me and everyone else?

More and more the students have to take over the perspective of those people who are involved in a specific moral conflict and have to examine their own moral decisions whether they can be accepted by other people and groups, too.

As an example of the structure of such sequences we'll point out three themes:

1. theme: "Discrimination against girls and women"
- sex specific socialisation in education, family, school... *(group)*
- sex specific discrimination in working life, working
  places, wages... *(society)*
- the role of girls and women in other
  cultures (Islam, India) *(Human Rights)*

2. theme: "Minorities"
- outsider in playing-groups, teams, peer-groups *(group)*
- discrimination in daily-life (immigrants,
  disabled children, unemployed...) *(society)*
- the suppression of political, social, cultural
  minorities *(Human Rights)*

3. theme; "Human dignity"
- "Do unto others as you would others do unto you!" Why
  I want to be respected as child, friend, pupil, student... *(group)*
- "That's degrading!" Unfair rules, degrading
  punishment, unjust laws... *(society)*
- Human Rights and the dignity of man/woman in
  our society, other cultures, religion... *(Human Rights)*

Each sequence of each theme tries to lead the pespective of students from an individual point of view to the next higher stage of moral-cognitive development.

From one issue-dilemma to the next they have to compare their individual perspective with that of their social group, the group perspective with that of our own society and the moral and legal basis of our society with the perspective of all other people.

We believe this procedure to be a promising approach to implement the moral-cognitive basis for Human Rights education.

1.6.6 NOTES AND BIBLIOGRAPHY

1. See RÜCKRIEM, G/M., Erziehung "zwischen" Pluralismus und Wertorientierung?, in: Demokratische Erziehung 6, 1980, pp. 398-494.
2. See NUNNER-WINKLER, G., Was bedeutet Kohlbergs Theorieansatz für die moderne bildungspolitische Diskussion in der BRD?. in: G. LIND, J. RASCHERT (ed.): Moralische Urteilsfähigkeit, Weinheim/- Basel 1987, pp. 17 seq.
3. see PIAGET, J., Das moralische Urteil beim Kinde, 1973, p. 117
4. see BROOKOVER, W., BEADY, C., FLOOD, P., SCHWEITZER, J., WISENBAKER, J., Social System and Student Achievement, Schools Can Make a Difference, Brooklyn, New York 1979, pp. 139. seq.
5. from: KOHLBERG, L., Moral Stages and Moralization; The Cognitive-Developmental Approach, in Moral Development and Behavior: Theory, Research and Social Issues, ed. Thomas Lickona (New York: Holt, Rinehart and Winston, 1976), pp. 34-35
6, KOHLBERG, L., WASSERMANN, E., RICHARDSON, N., Die Gerechte Schulkooperative. Ihre Theorie und das Experiment der Cambridge Cluster School, in: PORTELE, G. (ed.), Sozialisation und Moral, Weinheim/Basel 1978, p. 216.
7. Vgl. dzau DOBBELSTEIN, P., SCHIRP, H., Werteerziehung ind er Schule - aber wie?. Soest (Landesinstitut, issue No. 10) 1987.

1.7
Saying and Doing –
Why the Gap?

Jaques-André Tschoumy

Switzerland

1.7.1 SUMMARY

Why is there such a gap between saying and doing?
The theory put forward here is that human rights education covers a number of underlying issues which are rarely, if ever, made explicit, and that these issues stand in the way of such education which is characterised at present by its theoretical content and the limited extent to which it is practised in our schools.

We argue that the following types of obstacle are involved:
- educational;
- secular;
- philosophical;
- civil/civic.

Analysis of these issues would seem essential if we really want to free human rights education from the constraints which impede its introduction.

1.7.2 THE EDUCATIONAL OBSTACLE

The first obstacle is educational, since to teach human rights is to:

- try to convey knowledge which is both theoretical and practical;
- use teaching methods suited to this purpose;
- select teaching methods which centre on the pupil as active member of the community;
- devise an approach which is both universal and specific.

In short, it is to try and bring more democracy into schools, and this requires teacher training in new teaching methods.

Theoretical and practical knowledge

To teach human rights without referring to the texts is to run the risk of reducing human rights themselves to a vague, sentimental, moralistic and superficial system of beliefs, with no rational theory to support it. A theoretical approach to the texts in which these rights are set down is thus essential, however minimal and simplified that approach may have to be, because of the age of the children involved (simplified versions of the texts are available). This approach is all the more necessary for revealing the issues, and thus providing a link with the wider community and world in which the child must find a place.

This kind of teaching becomes, however, "a meal taken in a dream" (as the Chinese saying has it) if the ground is unprepared and it fails to connect with realities and experience that give it a meaning. Human rights are practised first of all in teaching that builds on real situations, on the classroom as living environment, where human relationships are made and remade day by day, and where personalities are formed. If law emerges from life to regulate and codify it, and to prevent the use of violence, then the first approach to law must be via the minor incidents of everyday life.

Theoretical and practical knowledge are thus linked objectives of human rights education, and specific instruction and general education are closely connected in this area.

This link is in itself an impediment. In fact, many teachers emphasise one of the approaches at the expense of the other and this choice militates against the desired advance of human rights in schools. Human rights education thus presupposes a choice of teaching methods, or special training or retraining.

This is a first obstacle.

Learning by questioning
In totalitarian and authoritarian states, education takes the form of deliberately feeding people answers. The civil societies which typify democratic states nurture the democratic instinct by asking questions. Indeed, it is only by constantly questioning its own values that society progresses towards greater justice; and it is by asking questions that human beings make social, scientific and cultural progress.

Many teachers' work still consists, however, of preparing children for the answers. Usually, they are even taught to answer questions which they have never actually asked themselves.
This second obstacle is a major one.

Teaching centred on the pupil as active member of the community
The practical teaching of human rights should, of course, centre on recognition of pupils as active members of a given community, in which they have acquired:

- a (recognised?) language;
- an (accepted?) culture;
- beliefs (myths?);
- a (stereotyped?) conception of the world;
- a (minority?) history.

This accumulated experience, and the theories concerning the world to which it gives rise, are the bases on which individuals explore their environment and seek to influence and change it. The teacher's job is to help them reformulate what is at first confused, encourage them to ask the right questions about the world around them, and guide them in reconsidering judgments which are flawed by prejudice, thus leading them to understand others and respect their right to be different.
But this approach is more time-consuming than others, and calls for almost total commitment.
These are further educational obstacles.

An approach which is both universal and specific
Every discipline is both specific and interdisciplinary.
In scope, human rights education is universal:
- it aims at universality (all human beings);
- it applies to all age-groups (without exception);
- it cuts across all disciplines;

- it questions everything that makes human beings human;
- it commits entire groups (and not parts of such groups)
(a college and not a teacher of human rights).

It must also, however, strike a balance between universal principles and local implications concerning:
- the age of the children;
- the cultures of the children;
- expectations;
- the political and social situation;
- curricula which are already overloaded;
- curricula which already embody various types of training for solidarity.

This striking of a balance between the universal and the particular must also be accompanied by openness to the entire educational community (parents, neighbourhoods, various milieus, etc.).
However schools, which first took institutional shape about 150 years ago, are often hermetically sealed off from the community.
This is another barrier.

More democracy in schools
Human rights can only be taught in school systems which themselves respect human rights.
This is why active pupil involvement is vital to human rights education.

In Geneva, in-service courses were organised which focused on real problems, and thus gave human rights a clear meaning for schools and projects. The collective decisions taken in schools were based on the personal experience of pupils and teachers, who geared their action to everyday realities.

But many school-users (parents) reject the notion of democratically run schools. Education systems thus pursue divergent aims.
This again is an obstacle.

Training
To teach human rights, teachers need special training, and this training is still lacking. This was one of the main conclusions reached at the Congress on human rights education, information and documentation, held in Malta in 1987, under the auspices of UNESCO.
This is a central issue.

The essential elements in such training are:

- familiarity with the international instruments and covenants binding on states, and with the institutions responsible for protecting human rights;
- resources (audio-visual, miscellaneous, data-processing, research findings);
- full and open contact between the various specialised fields and levels;
- protection:
  • human rights are not subversive, but the source of a new balance between public order and individual freedom;
  • human rights are not "left-wing", but reflect a collective need to live democratically.

All of this, however, takes time - and this is a further obstacle.

In short, human rights education set out to establish a new human order by:
- asking questions;
- engaging consciences and not simply intellects;
- centering on the human person;
- establishing links between areas which are usually sealed off or fragmented;
- reconciling a universal with a specific approach.

In view of all this, it is hardly surprising that the present situation leaves much to be desired.

1.7.3 THE OBSTACLE OF SECULARISM

Is human rights education an updated form of moral education? Does it typify secular education in the way that the Catholic mass - or moral instruction - once typified religious education?

The link is a close one, and may well form an impediment, if teachers feel that their job is merely to impart knowledge, and not to influence their pupils' conduct.

This view is mistaken for three reasons:

1 Teaching people how to live and giving them the right attitudes is vital at the end of this 20th century.

There are various reasons for this, and particularly:
- the need to come to terms with the spectacular increase in knowledge;
- the extent to which the various world communities have become interdependent.

This is a "challenge" which education congresses throughout the world have recognised.

2 To be secular is not to be neutral. Secularism did not mark rejection of the spiritual life, but rejection of the Church's hold on temporal affairs. This distinction is a vital one.

Secularism, which is bound to the state by a social contract obliging the latter to guarantee certain rights, is essentially an ethic, a concept defining the relationships between the individual and the community. This concept has broadened with the passing of time. To civil and political rights, the legacy of liberal thought, have been added economic, social and cultural rights, influenced by socialist thought. More recently, it has been suggested that rights to peace, to a healthy environment, to self-determination, to a heritage, to solidarity, should also be included. These are third generation human rights.

Human rights education can thus rely on the various factors which make this concept an original and rich one. As a product of our past, such education allows us to make a connection between politics and morality. Properly conducted and referring to clearly-identified situations, it will help us to avoid fussy formalism, empty moralising and any tendency to assume that behaviour is pre-determined.

3 Human rights ultimately derive, of course, from Christianity. But the connection is an ethical, rather than a religious one. Whether we like it or not, our Western society is Christian in its origins; from year's end to year's end, Christianity is woven into the pattern of our daily lives, and a very physical presence in all of our villages.

All the religions are divided today on the major issues of our time - on such controversial questions as ecumenism, liturgical reform and the laws governing birth, marriage and sexuality. These and all the other issues are no longer a source of division between churches, but of division within churches - creating a majority/minority split between the traditionalists, who want the church to distance itself from society, and the progressives, who want the church to live with society and meet it head-on.

This is the trend from which human rights ultimately derive. To this extent, moral issues have become part and parcel of our daily lives.

The history of human rights is thus a very recent one. Of course, its prehistory goes back to Socrates, who opposed the Athenian government;

to Hammurabi, who tried to define human nature; to Erasmus, who declared that war was folly; and to the English and Americans who drafted the first constitutions in the XVIIth and XVIIIth centuries. This "prehistoric" phase reached its climax on 26 August 1789, in the "Declaration of the Rights of Man and of the Citizen" - although the latter's bourgeois character was plain from the fact that it put property on the same footing as liberty and quality. After that, nothing happened for 150 years.

It was not until the middle of the XXth century that the real history of human rights began. The weakness of the democratic regimes between the wars led to a world war, which became a kind of crusade for human rights. A meeting between Roosevelt and Churchill in August 1941 led to the Atlantic Charter, proclaiming certain liberties; at the end of the war, the whole world learned of the horrors of the concentration camps; in 1946, the United Nations Economic and Social Council set up an 18-member Commission on Human Rights, which met for the first time in January 1947 and, after broad consultation and multiple hearings, proposed a Universal Declaration of Human Rights. A vote was taken on this declaration on 10 December 1948, with the following results:

- in favour: 48
- against: 0
- abstentions: 8

The countries which abstained were the USSR, the Ukraine, Bielo-Russia, Czechoslovakia, Poland, Yugoslavia, South Africa and Saudi Arabia. The articles concerning religious rights and freedom of expression were contrary to their national practices and prevented them from approving the text.
The Declaration itself was a typical product of the second half of the XXth century - century in which people ask themselves questions. Have you noticed how many book-titles are cast in question form?
All of these question-marks in bookshop windows are the signs of a century that is trying to find its way. The emergence of human rights is in every way consistent with this self-questioning tendency.

These questions forge a link between politics and morality, and this link is itself typical of our time. It seeks to uphold the concept of human dignity, and to raise this dignity to the status of a universal heritage.

For us Europeans, the prehistory and history of human rights have European roots - for us, this goes without saying.

In other parts of the world, it must be admitted that human rights often seem unduly "Eurocentric". But that is another story, and we do not intend to pursue it here.

1.7.4 THE PHILOSOPHICAL OBSTACLE

To teach human rights in schools is to touch on all the essential aspects of human life; to teach human rights in schools is to give priority to the concept of human dignity in every area of daily life; to teach human rights in schools is to seek the shared heritage of all cultures and countries; to teach human rights in schools is thus to strengthen the universality and indivisibility of those rights.

The interdisciplinary Centre for Ethics and Human Rights at the University of Fribourg devoted its 1985 colloquy to the indivisibility of human rights.

This approach envisages human beings:

- in all their aspects (cultural, social, political, economic, etc.) and;
- in all areas of their daily life: in fact, it is not only in far-off societies that the integrity of the human person is disregarded, but also in our own, highly technological societies - human rights can be flouted at work, on the bus, in the classroom, in the family, in the sports stadium.

Everyday experience shows us that life is structured in a way that compartmentalises it, whereas human rights aspire to be both universal and indivisible.

This philosophy is the reverse of that of Hobbes. For him, contractual, technical, artificial power is the only thing that counts. Comfort which is absolute and non-mediated must be sacrificed to comfort which is precarious and mediated. It thus becomes authority's task to provide this precarious, mediated comfort. In other words, happiness is mediated by the state, and other factors, such as dignity and solidarity, which might seem more important than state power, are entirely disregarded. For Hobbes, the right to security is paramount, and this makes him question the right to independence and the right to resist. In this view, any order - even an unjust one - is preferable to disorder. Anything else makes power insecure. The citizen must not, he feels, be torn between two rival types of authority. Spinoza considers, on the other hand, that power increases by being shared.

Paradoxically, this debate concerning the universality and indivisibility of human rights is itself divisive. And the resultant division is certainly one of the reasons for the slow introduction of human rights instruction in schools.

Schoolchildren, of course, are rarely involved in campaigning on the major public issues. Instead of going for the big abstract ideas, they naturally express their solidarity by working for specific causes. Action counts more than words. In one state, they mobilise in trying to locate missing persons or in opposing expulsions; in another, they are told that schools are expected to keep out of politics, and display no such solidarity.

Some countries try consciously to promote awareness among children and schoolchildren, allowing them to take the initiative and stand up for their own beliefs, and encouraging them to develop a critical spirit, based on values which are both personal and universal. The "whole" pupil is the focus - the universal, indivisible human being, living in a civil society.

In other countries, schools teach children what to think. They may teach them human rights, but rely on texts in doing so. The school in this case is a hermetic institution, and its task is fragmented by civic laws. Human rights here are neither universal nor indivisible.

"We go through life blindfold", as Kundera put it. When they were first formulated in 1789, human rights chiefly expressed the individual's desire for independence. "Everything for the individual in the community" was the catch-cry, and it implied that the human species fulfilled itself separately in every individual member of the community. This certainly appears to have been the main concern and guiding vision of the XVIIIth century humanists.

Since then, the growing importance of collective phenomena throughout the world has changed the situation profoundly. For a host of convergent reasons - the rapid development of ethnic, economic, political and psychological links - humanity is moving irreversibly towards the kind of "organo-psychic solidarity" envisaged by Teilhard de Chardin. Whether we like it or not, it is developing a new collective, overall identity on a world scale. Hence the inner conflict, affecting all of us today, between the personal element and the demands of the community, which are becoming even more pressing.

Looked at more closely, this conflict is merely an apparent one. Biologically, the human element is not self-sufficient. Everyone becomes a "person" dialectically, by interacting with others. Seen in this light, it at once becomes clear that the purpose of redefining human rights can no longer be, as it once was, to guarantee the individual maximum independence in the community, but to indicate how the inevitable process of collectivization can take place, not only without destroying, but actually enhancing the individual person.

The third generation human rights - the right to peace, to a healthy environment, to self-determination, to solidarity - are currently the subject of heated debate. It is not simply a matter of how to manage those rights. The argument is rooted in the different ways in which our contemporaries view the changes in the world around them - and the patterns of interdependence which they intend to establish.

When one thinks like this, in terms of changing ideas and attitudes, human rights are still young, at forty years old. Their progress can be measured neither in years, nor even decades. "It is very recently that the teaching of human rights has become a part of the study of international law in the law faculties of Latin America", Mr. Hector Gros Espiell (Uruguay), a judge in the International Court of Justice, told us in Geneva in July 1988.

Everything takes time, as the saying goes. Universality and indivisibility are challenges that we must meet every day.

1.7.5 THE CIVIC/CIVIL OBSTACLE

It is society, not the state, which gives rise to law. The state bottles the water, but the water comes from the spring, and not the state.

Totalitarian Sparta constantly fuelled discussion of this issue in Athens, where the community itself was the state - or better, where various communities provided forums for debate and so vitalised the state.

Democracy is obviously not born by magic. To teach human rights in schools is also to help defend the rights of civil society, to help make schools serve humanity by teaching pupils to respect life, show tolerance and overcome their selfishness, as the draft World Education Code puts it.

The obstacles, however, are numerous and varied.

Economic logic: The logic of the economists is utilitarian, not liberal. As Pareto says, effectiveness is the yardstick. Circumstances shape decisions, and it matters little if today's decision reverses yesterday's, provided that some benefit is expected and no harm is done elsewhere. Is economic logic compatible with human rights? In fact, it is neither incompatible nor compatible, but simply follows different lines. It is certain, however, that human beings count in this system only in terms of utility. And there are countless examples of an implacable economic logic that cares nothing for civil rights. But cannot the utilitarian and humanist outlooks come to terms?

Spatial logic: Traditionally, countries have always been organised on hierarchical lines. The powerful and the rich have always dominated the developed countries, and this has been accepted as completely natural.

Nowadays, all of this has changed. Something that was once a recognised privilege is now denounced as an injustice. Nowadays, people who can no longer influence events by voting simply "vote with their feet", moving from the city centres to outlying areas. This geographical mobility is paralleled by the growth of associations campaigning for social justice in the areas where social problems actually arise (parents' associations, neighbourhood associations, etc.). These associations set out to equalise advantages by introducing more flexible patterns of organisation. Regional planning is one way of meeting the demands of municipalities; the agreed delegation of certain powers is another way of meeting the demands of cantons. These flexible arrangements are no longer fixed and final, but shape the future by equalising advantages. Of course, they confuse the civic and political picture by establishing complex interlinkages which pose problems for planners. But the issue is a vital one: space must be redeployed by a gradual process of expansion or by joining up adjacent areas; these new zones must then be recognised and given the funds they need to play an active part in public life. Socio-economic classes have been thoroughly studied; increasingly, the emphasis will now shift to socio-spatial classes. We shall no longer think of communities that constitute areas, but of areas that constitute communities, making democracy increasingly effective at local level and so effecting the transition from a civic to a civil society, from a society of systems to a society of persons - a society of men and women who will make democracy a living reality by establishing a civil society and by holding out in defence of their own interests.

* * *

Creon and Antigone

This debate is the same as the debate in ancient Greece between Creon, who embodied the law of the "state", and Antigone, who embodied respect for "the good". Today, laws which violate human rights and legitimately be violated themselves, for the state always has certain conceptions of "rightness" in mind when it legislates to make rights effective in practice. The Swiss state, for example, legitimises its authority by referring to higher values, and particularly God and human rights.

There has been some slippage, however, and the state is increasingly becoming an end in itself; it is taking on an absolute character, particularly by unduly respecting the majority, while falling into the democratic temptation of failing to respect the minority sufficiently.

This closing-in of the system on itself leads to paralysis and conformity - and to disregarding of the individual. Resistance becomes a necessity, and indeed helps the state to grow, since the norm and its opposite, deviance, are different sides of the same coin. There can be no solidarity without a norm, but there is also "no innovation without the risk of deviance", as Professor Starobinski pointed out at the international seminar in Geneva on "Human rights: a vital form of deviance". In other words, liberty is nothing without the means of using it. The deviance which is part and parcel of a civil society is of major civic importance: it protects against the anarchist temptation of sweeping all the norms away, only to set other, more arbitrary norms in their place; above all, it protects against the totalitarian temptation of arguing that the norm is just because it is the same for everyone. "A lasting distrust of systems takes me to the threshold of liberty", Starobinski went on. "The threshold is certainly the grace needed, but it is not sufficient - it must be given practical expression", replied Bernard Béguin, who stressed the need for an independent press, serving civil societies above all, and for civil principles, underlying all national constitutions.

The right to resist is an old one. Was it not, after all, known in Athens? In fact, the finest flower of Athenian democracy was the "graphe para nomon", a law which served to check any temptation to abuse majority rule. Under this law, each and every citizen was free to question a majority decision. And this law found users in the midst of the most serious conflict Athens had ever known - the Peloponnesian War, which pitted it against Sparta. The political leadership in Athens had decided to sentence the citizens of Mytilene to death. At the request of various private citizens, the discussion was reopened under the law - the "graphe para nomon" - which that same political leadership had decreed. The earlier decision was then set aside.

Thus Athens not only introduced majority rule, obliging the minority to submit to the majority, but also took extreme care to preserve the civil communities (the minorities) and to invigorate the city's democratic and political life by having room for minority viewpoints. Does this approach make it harder for the state to function? Has the municipality of Lausanne not had recent, bitter experience of this in connection with the projected holding of the Olympic Games in Switzerland in 1996?

Is this model relevant to the protection of human rights today? Less than is generally claimed, and for various reasons:

In Vth century Athens, the Areopagus had lost its rights to a three-tier democratic structure:

- a democratic council, known as the Boule, with 500 members representing 35,000 citizens, chosen by lot every year from all citizens above the age of 30;
- a popular assembly of all the citizens of Athens, meeting 40 times a year; wielding absolute power, it decided on proposals put forward by the council;
- a people's tribunal, appointed by lot every year.

But to what extent were these rights really exercised? Democracy started to decline in the IVth century, and there are many texts hostile to democracy, particularly among the followers of Plato. The arguments are predictable:

- the authorities are blind;
- the state is governed by incompetents;
- the leaders are constantly yielding to the temptations of demagoguery;
- the popular assembly is patently apathetic (6,000 present out of 30,000, ie 20%).

Nor should it be forgotten that the city of Athens admitted only men to full citizenship (a woman could submit a petition only through a father or an uncle); defined only citizens' (and not human) rights; excluded slaves, who were often tortured to make them give evidence; excluded foreigners and people of other races, who paid taxes and served in the army, but had no landed property or political rights; shut itself off from the outside world, while forcing allied cities to pay a tribute on which its own prosperity was based: its fleet and its Parthenon were in fact the "gifts" of exploited and excluded allies.

And so Greek democracy, taken as a whole, is scarcely a fruitful model for human rights today.

But Athenian democracy did achieve a number of important breakthroughs:

- *In its three-tier democratic structure*, as we have seen. Important issues were discussed in various forums, and this was a major source of democracy.
- *In the "graphe para nomon" (the law protecting minorities)*. This was the noblest product of Athenian democracy, serving as a shield against bad majority decisions. Under this law, any citizen could contest a majority decision. During the Peloponnesian War between Athens and Sparta, for example, a decision had been taken to put the citizens of Mytilene to death. The discussion was reopened at the request of various citizens and the decision itself was revoked. Majority rule, obliging the minority to accept the majority's views, is an Athenian achievement, but the rights of minorities must also be carefully protected. This is why the minority always needs a "graphe para nomon" - giving civil societies within the state the right to oppose government decisions. Of course, this complicates the workings of the state - but is it not the sign of a smoothly functioning democracy?
- *In having no censorship*. During the Peloponnesian War, Aristophanes held the Athenian leadership and the war itself up to ridicule, bringing on stage an Athenian peasant, who made peace with Sparta single-handed. Aristophanes could say anything he liked without fear of censorship. This freedom is all the more surprising when one remembers that his plays were aimed at the general public.

Plato, too, was able to teach without hindrance. But between the two, Socrates was brought to trial because he had questioned the value of democracy, criticised the leadership and failed to conceal his admiration for Sparta. It is true that he was accused of corrupting the young, and that his trial turned chiefly on ethical and religious issues. But it was, above all, an assault on freedom of thought, conduct and speech.

Plato, Socrates, Aristophanes - three aspects of a far from simple democracy.

To establish a democratic state is certainly to provide protection against anarchy and totalitarianism; but this is not enough. To establish a democratic state is also to build shock-absorbers and antibodies into the system itself; strong civil societies and a new local democracy are what is needed to vitalise modern democracy.

To teach human rights in schools is to strengthen the protection of the rights of civil societies within democratic systems and thus to maintain a constant dialectic. Human rights, as we see, never give us a chance to relax.

Nowadays, every man and woman can rely on a whole battery of international reference texts, which allow him or her to determine whether the legal or de facto authorities are respecting the various human rights declarations and conventions. Democracy goes hand in hand with this permanent right to comparison, information and vigilance. Vigilance allows the individual citizen to ensure that rights are respected in practice. If disagreements arise, he has both the right and the duty to resist. The state is subject to certain higher laws, and its own health depends on the civil health of the community. Putting Socrates on trial is always an assault on democracy. But it is also a part of democracy, for democracy is something one does not live blindfold.

Is this analysis in line with the expectations of our civic societies?

It is, at any rate, an analysis that is often heard nowadays, and major books are devoted to it; daily life obliges us to look squarely at the relationship between the civic and the civil.
Our school systems must not evade this issue.

1.7.6 BY WAY OF CONCLUSION

A product of our recent history, and thus very recent themselves, human rights raise major and controversial issues of equity and justice: individual equity in 1789, social equity later, and now the balance between individual and community interests. They make it possible to link politics with morality.

Paradoxically, the debate on the universality and indivisibility of rights is itself divisive. It may no longer be made light of - which is progress in itself - but it still gives rise to more talk than action: that resistance persists, is certain.

It was analysis of these various types of resistance which interested us, and particularly the various practical ways in which they affect the teaching of human rights. Admittedly, there are already many signs that the teaching of human rights is progressing, both in the Swiss cantons and in other countries. But it must be confessed that this progress in still very modest. Is this mere chance? Surely not. There are educational, secular, philosophical and political impediments. And they are so powerful that human rights and the teaching of human rights are already declining in certain countries of Latin America.

And so vigilance is needed. The fact that every human beings have certain rights which they share with all other human beings; that they know that they have them; that they know, above all, that they can claim them, wherever they are, in the face of high-handedness and despotism - all of this is one of the great achievements of our century.

The fact that our civil societies all refer to one code is of paramount civic significance: it protects us against the temptations of anarchy, which sweeps away all the norms, and sets other, more arbitrary norms in their place; above all, it protects us against the temptations of totalitarianism, which tries to convince us that the norm is just, simply because it is the same for everyone.

Some of the greatest problems of our time could be solved if we realigned teaching practice and lesson content to provide a genuine human rights education. To do this would be to steer a new course - to give today's schools a new ethical code.

1.8

Socialization and Human Rights Education – The Example of France

François Audigier

France

1.8.1 SUMMARY

For generations, men have been endeavouring to hand down to succeeding generations the rules governing life in society and human relations. School has come to play an essential role in the transmission of these rules, in whose definition human rights have introduced a radically new concept. Human rights education must be based on the most significant elements of this new concept, which may be summed up by a formula applicable at all levels of society: "Humanity is the source and subject of the law".

This paper comprises three aspects:
- a rapid introduction to some of the main human rights issues, that is to say the new elements contributing to the way human relations are to be lived out and conceived. By taking these elements as a reference, we can construct a form of human rights education which is neither constrictive nor normative but designed, on the contrary, to educate for freedom;
- the problems and difficulties which any educationalist or team of educationalists working on a human rights education scheme will have to cope with. These problems are intrinsically linked to the very subject

of human rights. By bringing them to light, we can avoid unnecessary controversies and forestall certain setbacks;
- the various components of a human rights education scheme in which the teaching of factual knowledge is combined with its practical application, taking life in school as its primary location.

1.8.2 INTRODUCTION

Human rights education[1] is a fashionable theme. Over and above the real or supposed effects of the Bicentenary, it is a subject, the study and application of which survives social fluctuations and fashions. Developments in a number of countries provide positive, but also, alas, sometimes negative illustrations of the permanent relevance of HR. In the field of education alone, which is the main subject of this colloquy, numerous initiatives, experiments and debates have been conducted in recent decades in our various countries and will continue to be so in future. At all events, the theme is inexhaustible and the very principles of HR are undergoing constant exploration and discussion. The debate around HR is a healthy sign; I personally dread the day when they are accepted by all without comment other than conventional words of agreement. My paper, essentially based on experience acquired in the course of an educational experiment involving some hundred elementary, lower secondary and secondary school teachers between 1984 and June 1987, will be divided into three sections:

- an investigation into the new elements which HR contribute to the school and the justification for teaching them;
- an analysis of some of the problems which will always beset HRE and which I shall divide into three areas:
  • the respective weight to be given to what is "private" and what is "public", namely the difficult distinction between that which pertains to the actions of individuals and their private sphere and that which can or must be the responsibility of society as a whole;
  • attitudes to social roles, that is to say the way individuals perceive their place in society;
  • the relationship between educational and social aims, given that HRE, like other "fields" such as moral or civic education, explicitly seeks to train children for life in society both today and in the future;
- the various components of HRE, some of which concern the imparting of knowledge and others life in school. I shall be basing my arguments on

[1] The abbreviations HRE and HR will henceforth be used to designate human rights education and human rigths respectively.

the principle of the imperative need for a solid acquisition of theory and matching practical application focusing mainly on the school environment.

The purpose of my paper is certainly not to provide cut-and-dried answers to questions which can be resolved once and for all, since the answers must be worked out as part of the educational process itself and the flow of questions is unending. It is my hope to spark off as rich a debate as possible on both the way in which certain problems should be formulated and the outline answers it is possible to trace.

1.8.3 WHAT NEW CONTRIBUTION DO HUMAN RIGHTS MAKE AND WHAT IS THE JUSTIFICATION FOR TEACHING THEM?

From generation to generation, human beings have tried to hand down knowledge of the world, rules of individual behaviour and rules of collective living, that is to say principles governing relations between human beings, and between human beings and authorities, which provide the basis for settlement of conflicts. These three elements to be handed down could be expressed in today's terminology as teaching, upbringing and socialisation.
Virtually throughout history and in most civilisations, the family and social groups took responsibility for this process; the school then appeared on the scene and more recently, we can add the media and the peer group as agents of change.

To teach, bring up and socialise is to pass on not only knowledge and positive learning, but also values, moral precepts and judgments on right and wrong. This is not something which began with the advent of human rights! This observation provides us at the outset with one requirement of HRE. Let us not confuse one thing with another: to teach a pupil not to "pinch" his neighbour's belongings is indeed indispensable from the point of view of the rules of life in society, but it is not human rights education (although there is a connection). The question, then, is: what does the HR field have to say that is new, different and "revolutionary"? What are the main points to be emphasised? It is on the basis of these main points that we shall be able to define, promote and reflect upon human rights education with the demands and difficulties in entails.

Without entering into the substance of HR, their history and the legal principles that they imply, I would affirm that there can be no coherent HRE worthy of the name which does not take as its starting point a serious examination of the essential contributions HR have made to human and social relations. Consequently, we must begin our examination of HRE with an analysis of these contributions. Each one will be set out very briefly in this paper, but I am fully aware that each of them would warrant a long discussion:

- *Humanity is the source and subject of the law.* This simple but eloquent formula constitutes a complete reversal in relation to former ideas on the origin of law; it is far from enjoying unanimous support in all parts of the world; it simply means that the law develops in free discussion between all people and that humanity is the purpose of the law;
- *the safeguarding of human rights is linked of the participation of everyone in the formulation of the law,* that is to say in the exercise of *power*; HR are thus intrinsically bound up with *democracy*, whatever its institutional form; this assertion extends to the multiplicity of rules governing everyday and occupational life; power must extend to all levels of the community;
- *respect for my rights is safeguarded by my respect for the rights of others*; there is *equality* of all before the law and *reciprocity* of this law; this leads to the "rights presuppose duties" maxim and the necessity of solidarity;
- the foregoing principles imply settlement of conflicts *without annihilation of the adversary*;
- *civil and political rights and economic and social rights are complementary*; this relationship can be understood in connection with the concept of *dignity* which appears in Article 1 of the Universal Declaration. On this difficult and controversial subject, I would simply point out that the exercise of freedoms requires a minimum degree of access to decent living conditions, culture and information; however, the other side of the coin is more difficult to grasp. Economic and social rights imply intervention by the state and their satisfaction is linked to each country's development and the need for arbitration between rights and aspirations which may appear to be contradictory; but it is precisely because there is a need for arbitration that it is impossible to satisfy these economic and social rights without civil and real political rights; the latter are a necessary condition for the participation of each citizen in the choices made and in the arbitration processes conducted with a view to satisfying rights which concern him. Where these rights and freedoms are not respected, who will be the arbiter? Who will choose which of the man rights and claims deserve to take priority? A group, a clan, a party? In the final analysis, an indeterminate number of individuals who take upon themselves the power to decide on the behalf of others ...;
- the *delimitation of the scope of the law*, that is to say the scope of state power, the extent as limitation of the powers of an authority or the

authorities; HR say that individuals, all human beings, have *rights to protect them against power*, against abuses of power;
- finally, let us bear in mind that *the law states the rule and not the facts* as they occur in the day-to-day life of our society; the law does not describe reality; this being so, HR are always before and not behind us, but being ahead of us, they are of necessity open-ended, laying down ethical principles to be used in devising and organising the future, but leaving society free to discuss and choose its own future options depending on the interplay of human forces.

To these points of reference, I would add that HR can only be understood if they are seen as "congenitally" linked to a whole series of words and concepts: freedom, equality, legislation, law, state, property, individual, nation, dignity, etc. without which they cannot be perceived or conceived. This gives us an eminently complex body of references and it is important that we should grasp well the main relationships within it.

This "open-endedness" of HR leads us to think of their enshrinement in positive law as at once desirable at all times and constantly in need of review; HR are also a tool for criticism of social situations on the basis of an *"ethical reference"* whose unpredetermined character prevents it from being normative.

HRE cannot therefore be reduced to this set of principles. Neither can it be dissociated from them, and only if it is based on solid knowledge of these rights, their implementation and their ethical and political implications can it develop, without being normative nor prescriptive, as a forward-looking process deeply associated with education for freedom.

However, once these statements have been made and these questions raised, there is one more question which we must ask ourselves, namely: what are teachers' credentials and justification for carrying out HRE? Teachers "know things', and it is on the strength of this knowledge that they are present in the school and that society entrusts to them the task of teaching its young people. But we have seen that the teacher also has a role to play in children's upbringing and socialisation; in our French school tradition, these two latter aspects are subject to debate and where the school genuinely applies them, a problem area emerges which should be neither disregarded nor underestimated. To reject this dimension of the school is to deceive oneself as to the teacher's real role and reject the function of the school as an institution within society, while to accept it is to open the door to a host of questionings and debates.

The teacher's justification for teaching human rights is also based on the fact that we live in societies which proclaim human rights, and which therefore subject their systems of law to this ethical vision of social relations. Human rights are at the root of positive law in at least two respects:

- The countries present here have virtually all signed the international instruments which make up the world charter of human rights, namely the Universal Declaration and the two Covenants of 1966 together with the European Convention. I would add that in the case of France the preamble of the current constitution echoes the preamble of the constitution of the Fourth Republic, itself built around the 1789 Declaration, supplemented by a statement of economic and social rights. These texts are the reference upon which the Constitutional Court bases its judgments on the constitutionality of acts of legislation.
- Teachers are also citizens and in this capacity they play a part in handing down these rules of community living to succeeding generations. Their institutional position vests them with particular responsibility in this connection.

1.8.4 HUMAN RIGHTS EDUCATION ALSO MEANS FACING A NUMBER OF PROBLEMS WHICH WE CANNOT AVOID

In the first part of this paper, I identified a number of requirements and principles deriving directly from close consideration of what human rights stand for and what factual knowledge is necessary to their understanding. But introducing HRE does not follow mechanically from these principles and this substance; their incorporation into the reality of life in school and society shows up and brings into play a whole series of factors whose existence and relative weight must be carefully considered.

Accordingly, before putting forward a number of principles which may serve towards the construction of an HRE scheme and defining more precisely the foundations on which to build this education in the classroom, I shall analyse a number of these factors, particularly those which are indicative of the difficulties which have to be faced. Educating children to appreciate human rights is not a neutral operation; teaching multiplication or the names of the capitals of the world's principal countries[2] is not the same thing as teaching HR or educating children in HR. I shall now look

[2] These comparisons do not mean that teaching multiplication or even capital cities is an easy and straightforward operation; as an educational theorist, I am all too aware of the difficulties of any form of learning, including that of apparently simple matters. This reference is merely intended to show the specific characteristics of HR in relation to other more usual school subjects.

into various aspects of the special nature of HRE, as I believe that any teacher, any team of teachers or educationists proposing to construct and implement a scheme of this kind should be aware of these particularities and that the effectiveness of such a scheme depends on making these particularities an important dimension or even the backbone of the scheme itself. At all events, not to acknowledge them is to court disappointment or even failure.

I shall mention three of these specific characteristics. For the purposes of these remarks, I shall include together with HRE the adjoining disciplines or "fields" of civics, politics and ethics. I shall also take a few brief looks at the history of the teaching of these disciplines (or fields), inasmuch as the revelation of certain constant factors or breaks with tradition may be very illuminating for our understanding of the situation today.

1.8.4.1 The respective roles of the private and public spheres in the training of young people

What are the respective roles played in the education of young people by the various agents I identified in my introduction, namely school, the family, the media and the peer group?

Behind this question, there appears another: to whom does the child belong? To himself, is generally the reply in respectable circles, but I very much fear that this expression of consensus will help us but little in our observations! The child or young person will always be the object of attempts to exert power and influence; the crux of the matter is the form taken by this power and the end which it pursues, the extent to which it is compatible with a genuine education for freedom.

This question of belonging, which is ultimately a question of responsibility, leads on to two further questions:

- what is the *aim* of this education?
- what is the *basis* of this education?

To address the question of the aim, I shall simply refer to the contents of the official statements and texts according to which this education is organised. For over a century, their contents have been oscillating between two positions: inculcating obedience or developing a sense of criticism. Should we train citizens who are free and therefore critical[3] or citizens who obey laws and observe social standards? Admittedly, it is possible to obey

---

[3] Let me point out straight away that by criticism I mean the rational examination of any proposal and not the exercise of unfounded opposition.

the law *and* have a critical mind, but handing down the social norm and training the critical faculties, with the risk for the future this implies, represent two options which are sometimes at odds.

The question of the basis raises that of values; I would state at the outset that there is no education which is not based on values. For thousands of years it was religion which provided this reference and in many societies it still does so today. But HR offers another reference. It is not my task here to deal with relations between religions and HR, which would at all events be beyond my competence; but to do so, I think it would be necessary to refer to my earlier remarks on the fundamental contributions made by HR.

Historically speaking, however, it may be said that, in our French tradition, HR and religion were long in opposition to one another. Today, although these rights enjoy some degree of consensus as an ethical reference which we apply to our perception of human relations in society, one question remains very pressing and wide open, namely the expression of the right to be different, in the context of universalism. This is not an abstract question and many teachers and educationalists experience it in their day-to-day professional lives, particularly in multicultural classes.

The respective roles of the various agents and the aim and basis of this education are concerns which may be reactivated by conflicts between those exercising authority over young people and between young people and these authorities. To promote human rights education is to introduce into the very school environment the question of authorities, their legitimacy and the way in which they are exercised.

1.8.4.2 Relationships between academic and social aims

What is the use of what is taught in school? This question, which may appear thoroughly trivial, will receive contradictory replies depending on whether heed is paid to the (self-justificatory) statements of adults or the way in which the system operates; adults, particularly those who make their living from the school, generally assert that what is taught is of use "for real life", while in the practice of the institution, what is taught in school serves first of all to solve school problems, problems which are mostly well defined and which, whatever teachers may expect of them, encourage primarily the reproduction of models.

But what, then, is the role of HRE in this contradiction? With what school problems can HRE be associated? Is it merely a matter of knowledge? In that case, there are oral or written examinations, the traditional forms of assessment. But HRE professes an inherently social purpose; a successful form of HRE is one which leads to behaviour in keeping with the values associated with human rights.

What can be the use of a human rights education restricted to facts, to more or less formal data? The aim pursued, stated and professed is indeed to influence behaviour, namely the way people live out their relations with other individuals and with society. It is therefore an attempt to inform and develop behaviour consonant with certain values; but if that is indeed the aim of HRE, what are these forms of behaviour? Are we reduced to being able to designate only the kind of behaviour which is not consonant with the values in question? And what does assessing behaviour involve? How? Which? Where? When? Must we remain within the confines of the classroom? Those of the school? Should we take an interest in what happens outside school? And do we even have the right to do so? We may be judges of knowledge; there is some reference for that; we may know by what yardstick to measure a statement of knowledge, but what about behaviour?!!! We come up against a formidable ethical question which confronts us again with the problems of separation of the private and public spheres and problems of authority; a formidable ethical question inasmuch as adherence to the ethics of HR is quintessentially an act of freedom.

The three tasks assigned to the school by society, namely teaching, upbringing and socialisation, are situated at very different levels. When we are dealing with the learning of a particular mathematical operation or physical law, a rule of grammar, or an historical event, we are essentially in the teaching mode, although a precise analysis would show the importance of the individual who is learning; but when we place ourselves within the context of HRE, the situation is profoundly and directly different; we are involved in teaching, education and socialisation all at once. We can try to master the teaching aspect, but the other two aspects are much more difficult to grasp.

This difficulty is aggravated by the fact that, in this field, the school lives on a kind of illusion which is its very raison d'être: an increase in positive knowledge is supposed to produce more rational behaviour and, conversely, knowledge of situations in which it can be said that human rights are experienced is supposed to make it easier to learn the principles concerned; there is thus supposed to be a fairly simple relationship between "concrete" matters and other more abstract ones. This is entirely unproven; knowledge of the laws of deformation of bodies by speed does not necessarily make the physicist more respectful of speed limits on the highway than someone who is ignorant of these laws! However, conversely, we may comfort ourselves with the thought that if HR are to be respected, it is preferable that they should be known!

1.8.4.3 Pupils' representations

I have just described HRE as representing, by virtue of its very subject matter, three dimensions named as teaching, education and socialisation. I shall further expand my argument by looking into what is so special about this subject matter. To educate children in HR is not only to teach them the definition of the terms "human" and "rights"! Although that in itself is a necessary and not altogether easy task! However, as we have seen, it does mean enabling them to handle a whole range of concepts such as freedom, equality, law, sovereignty, power, individual, nation, etc.

This process is a complex one because these terms are themselves complex; it is never finished because none of these concepts can be encased in a cut-and-dried and final definition. But for us, as educationalists, this complexity has another dimension; each of these concepts and their interrelations confront and directly challenge some of the pupils' representations[4].

Constructing these concepts means giving a functional meaning to these words in a precise theoretical context, a meaning which claims to be distinct from the common sense meaning while aspiring to take its place; this in turn is tantamount to challenging the way in which individuals perceive society and human relations and the way he sees their own place in society and their relations with others.

To educate people in human rights is not a matter of teaching knowledge limited to technical matters and relating to a more or less self-contained category of problems (if such a thing exists); rather, it calls into question everyone's innermost conscience. HRE is based on knowledge which I would refer to as "hot", that is to say knowledge which is itself the object of ideological and political debates, knowledge which is at once made up of positive cognitions and systems of values. No one will ever be able to enclose freedom in one formal definition which comprises all the meanings of the word. Every individual has an idea, a personal understanding of freedom. The teacher or tutor collides head on with these representations;

[4] I shall use the term representations here in the meaning given to it by social psychology, a meaning associated with terms such as conception, mental image, practical theory, etc. A representation is a complex set of cognitions, opinions and attitudes on a given subject.

they resist and hence the ideas which the teacher[5] is trying to construct are often opposed to the "common sense" of everyone's views of the pupils, individually and collectively. These representations put up a resistance all the stronger since everyone needs them to live; they are systems of thought which enable us to grasp the world, act in it and incorporate new information. The task of the person teaching HR is made all the more difficult since the words to construct meaning within the HR context are the same words that are used in day-to-day speech.

This question of representations also raises the whole problem of the relationship between the irrational and the rational; that is, the influence or effect of rational arguments on attitudes, judgments and forms of behaviour which are themselves not necessarily rational. The most common example is that of racism. How effective are scientific arguments in this connection as a means of combating judgements which do not have a rational source? The weight of moral arguments is also questionable. Proving to a pupil that racism is devoid of any biological basis will not prevent him from using a racist argument if he is attacked by coloured people or people of an appearance different from his own when travelling back to his suburban tower block. To say as much is by no means to question the value in school of a scientific explanation of the absence of any rational basis for racism, but rather to prompt us to be clearly aware of what is at stake and of our own possibilities as teachers. Education for human rights is a complex and difficult undertaking which brings many equally complex elements into play; teachers should be warned and made aware of this so that they do not select the wrong targets or cherish vain illusions and so that they know where they are going and where the bounds of the possible lie.

1.8.5 THE VARIOUS COMPONENTS OF HUMAN RIGHTS EDUCATION

By studying the intrinsic requirements for an HR syllabus and some of the difficulties associated with this form of education, a number of efficient instruments may be acquired with which to construct an HRE scheme along more practical and more effective lines. I shall now look into several ingredients of any scheme of this kind. To be consistent with the subject matter of HR and at the same time respect the aims of this education, it is necessary to work in two dimensions: a cognitive dimension providing HRE theory and a practical dimension placing it in the context of day-to-day life. The latter will take as its almost exclusive framework the school environment itself. The sole purpose of this restriction is to indicate the

[5] Teachers, too, have their representations, a point to which I shall return.

high degree of prudence called for in what I have termed the relationship or opposition between the public and private spheres. These two dimensions are themselves made up of various components and encompass the various aspects which a more or less ideally constructed HRE scheme should take into account.

1.8.5.1 To educate for human rights is to impart knowledge based on three components

- Firstly, *the formation of a set of concepts* which take on particular significance in the field of HR: freedom, equality, dignity, sovereignty, law, etc. The acquisition of these concepts, a necessarily slow process which can never be fully completed, touches on all school disciplines. There is, of course, a "hard core" made up of those disciplines in which humanity is the object of study: history, civic education, geography, economic and social sciences, philosophy, literature, biology, etc. Other subjects have a more distant relationship to HRE, but all teaching is also a social practice, and as this applies to all disciplines, they shall be taken into account once more in the second dimension of HRE pertaining to praxis. When I say that all disciplines may be concerned, that implies that the ideal scheme is one based on a multidisciplinary team of teachers, and introduced from elementary school onwards.
- Secondly, *discussion of the foundations of human rights and the questions to which they give rise today*: this should cover historical and philosophical aspects (although I know that this situation is peculiar to France), ethical issues and the new questions raised by developments in the media, data processing and progress in biological sciences. These topics will fall more within the scope of history and philosophy teachers, but will also concern those who teach civic education (where they are not also responsible for history); similarly, the biology teacher is best placed to deal with the biological issues; there too, however, there is much to be gained from a multidisciplinary approach. Many important discussions taking place today are at the outer limit of the traditional academic disciplines.
- Lastly, *legal training* providing everyone with a grasp of his minimum rights, the rules of community living and ways and means of resolving conflicts ... as well as providing insight into how the law is applied and how justice works. The French school situation is particularly lacking in this respect as law is not a school discipline, nor is it desirable that it should become one. However, an element of law is introduced into teaching syllabuses, particularly through civics education. There is therefore a need for a systematic enquiry - which to my knowledge has never taken place - into ways of incorporating law into the subjects on

the curriculum which concern contemporary society at large in its legal, civic and political aspects. The other problem is that of teacher training, initial and in-service training, as those who teach civics in France at the present time are badly equipped in this respect.

I shall conclude my remarks on the knowledge aspect with a series sets of questions;

- *To what specific practices in school* can HRE be related? This is a vast question which I shall address from an historical point of view, emphasising first of all the continuing hesitation of the education authorities as to whether or not civics should be made an autonomous discipline. Civic education and HRE concern a priori everyone, that is to say all those who play an active part in the education system; but to content oneself with a statement of this kind is to take the risk, which has been pointed out on innumerable occasions, of diluting this education; since it is the concern of everyone, no one really takes it in hand! But to make it a separate discipline is to take the opposite risk of allowing this form of education-socialisation to be reduced to 1 or 2 hours a week and to become the preserve of a specialist teacher! A study of the texts issued by the French authorities over more than a century reveals constant hesitations, which are the sign of constant dissatisfaction. But let us continue our enquiry as to whether or not HRE should be a separate discipline and ask ourselves what determines the lasting establishment of an academic discipline in school. Apart from external conditions, for instance social demand, historic circumstances[6] etc., I shall point out three elements which enable a school discipline to survive:
  - a "vulgate", that is to say a corpus of knowledge more or less accepted as characteristic of the discipline in question and whose validity is attested by all those associated with schooling;
  - a standard method which enables the organisation of a lesson or a course to be described in formal terms;
  - an assessment procedure.

Let us now relate these three elements to civic education and HRE:

  - as far as the "vulgate" is concerned, their subject matter is unstable, debatable and open to social and ideological controversy; the constitution of a minimum body of knowledge accepted by all is

[6] It is interesting to note that the question of civic education arises at times of crisis, in periods when society is seeking to assert its identity around common values.

constantly called into question. One way out is to draw up an organisational chart, which has the appearance of objectivity;

- as regards methods, there has been a constant fluctuation between: the "morals lesson" (this has sometimes worked, particularly with the youngest age groups); the raw organisational chart with the people at the bottom, the authorities at the top and, if possible, a pyramid shape (but what meaning can these very abstract forms of organisation have for young people who as yet enjoy no political rights?); surveys based on the active method (but there are many indications that, once the active phase was completed, the synthesis and learning phases left mutch to be desired);
- as regards assessment, I would repeat my previous observations; there is nothing wrong with examinations, but they are a pale shadow of what HRE seeks to achieve, namely knowledge applicable to real life!
- The question of representations. I return here from a more practical angle to this vast question which I have already touched upon, in order to demonstrate some of its aspects. Teachers have representations just as pupils do. It is important to work on those of teachers undergoing training, a frequent practice in the training of adults.

As far as children are concerned, I would cite a number of examples. Some representations are the products of confusion: for instance that all the inhabitants of a country are citizens; that the law is just and durable; that democracy is equivalent to, or merely a matter of, the right to vote; that arbitrariness is the equivalent of totalitarianism and that nationalism is the same as xenophobia.

Others are to be interpreted rather as over-simplifications: freedom is the power to do whatever you like; equality is first of all to be understood as social equality; pupils have considerable difficulties in conceiving legal or political equality and this leads to mental blocks when it comes to understanding, for instance, what was at stake in the French Revolution and what were its achievements.

These brief notes are intended to emphasise the importance of representations and demonstrate the need for the teacher to take them into account. It would be idle to hope to destroy them (in any case, ethically speaking, do we have the right to do so?). In any representation, there is a mixture of knowledge and opinions. We can act on the knowledge component and show whether it is false, approximate, devoid of true bases, etc., that it is an unsubstantiated belief and not the product of a rational examination. For instance, not everyone who lives in a country is necessarily a citizen of that country, democracy cannot be reduced to the formal aspect of the right to vote, etc. The teacher can work on this dimension and, in particular, at the interface between knowledge and opinions, in order to help the pupils to distinguish the two and teach them to sort one from the other. An opinion, however, is not bad or false, it is what it is, an expression in which knowledge is mingled with judgments which are frequently emotional; it is important to subject opinion to doubt and criticism. In this respect, work on representations is useful and indeed necessary, not to destroy them but to put expressions of opinion in their correct perspective in relation to expressions of knowledge, in full awareness, of course, that in matters of HR as in all social sciences, it is virtually impossible to disentangle the two; however, conversely, we do not have the right to say anything we like on any subject we choose! The requirement of truth is intrinsic to freedom of debate.

To conclude my observations on knowledge, I shall mention a *last series of difficulties*: it is not always possible to address all issues with pupils, particularly the youngest. There are no taboo subjects, but there are subjects whose emotional implications are such that their study or presence in class is problematic; these subjects pose once more the problem of public and private spheres. Among these subjects are: torture, delinquency, rape, sexuality. To emphasise the force of these subjects is not to outlaw their presence in HRE, but to invite teachers to exercise the greatest possible prudence in dealing with them should they do so.

1.8.5.2 HRE means challenging and transforming school life

The problems which seem to me to attend HRE work on school life are as follows: if the school institution wishes to pass on values, it must accept that its practices and operation should be subject to criticism based on the very values it claims to pass on; this criticism is exercised in respect for the purpose of the school, which is to help to promote a fundamental right, the right to upbringing and education.

Having postulated this principle, I would also affirm that the school is not a microcosm of society; it is a social institution with its actors and their interrelations. Its purpose is not to ape grown-up society by transplanting adult activities into its confines.

HR give meaning, purpose and substance to a whole series of practices relating to: class delegates, access to information in school - particularly as regards careers guidance, the provision of facilities for free speech, in-house rules and, in the case of elementary school, class regulations or co-operative classes.

It is thus possible to indicate three potential components for this dimension of school life:

- *within the class itself*: though I in no way wish to denigrate teachers, it is obvious that, from the point of view of HR values, life in the classroom offers potential topics for inquiries and initiatives: the clarity of the didactic contract, the pupils' opportunities for expression, conflict management, relations between adults and young people, etc. The idea is not to transform the classroom into a permanent forum, but it is important that moments and places for free speech should be offered and used; many experiments have shown the effectiveness and value of initiatives along these lines;

- *within the framework of the school establishment*: there is a change of scale here and pupils will not so easily have the opportunity to put their points of view directly. It is here that the idea of delegation or representation comes into its own. I would insist that it is perfectly natural for pupils not to know how to speak in public or how to use this delegation mechanism effectively. These representative institutions must be accompanied by corresponding training, although for us as adults the situation is terribly wearing, as the process has to be begun over again with each generation of pupils, whereas we would so like to be able ourselves to progress with the pupils. But let us look around us and observe modestly how the delegation mechanisms in our adult society operate and have a little patience and indulgence for young people's capacity to imitate us, and patience and hope as regards their capacity not to imitate us, not to reproduce the spectacle with which we too often present them!

- *the extra-curricular sphere*: by this I mean a whole series of activities which exist within the school institution without having the compulsory nature of lessons: clubs and societies, activities in youth clubs or centres, educational projects[7] etc. A whole range of possibilities opens up here for individual initiatives freed from the constraints of the system; thus, PAEs or clubs on the third world, human rights, etc., may often be found, many of which come to grips with current events or environmental issues. These activities may be the occasion for activities directed towards the world outside the school, or towards other pupils, by means of exhibitions, opinion polls, etc. These opportunities for extra-curricular projects are frequently exploited and give rise to valuable and effective initiatives.

[7] In France, provision is made for "projets d'action éducative" (PAE), namely schemes prepared for the benefit of pupils by teams of teachers and possibly including other categories of staff: they consist of activities not directly concerned with normal lessons and both adults and young people participate on a voluntary basis. For instance, an educational project on human rights could be set up.

Finally, it must be borne in mind that all these activities affecting school life require, even more than do those concerning acquisition of knowledge, a positive attitude on the part of school administrations and particularly head teachers.

1.8.6 CONCLUSION

In conclusion, I shall recapitulate a number of elements to be taken into account in any HRE scheme, and would beg the reader to excuse any repetitions. In so doing, I shall concentrate on oppositions which should be treated as integral parts of the scheme, which are complementary and not alternatives; HRE constantly refers us to new and frequently contradictory aspects, and taking these into account is an important way of ensuring that this education does not degenerate into the formulation of rules or postulates.

HRE must be based on the rigorous and progressive acquisition of knowledge, the application of that knowledge and the testing of its effectiveness, primarily within the confines of school life. The point of this combination of theory and practice is to avoid the over-abstract treatment of knowledge which is often complex, and the over-simplification inherent in any once-only consideration of specific examples or personal experience.

It is necessary to *study not historical situations* for which there exist reasonably reliable approaches and interpretations, references and arguments, but also *contemporary situations*, for instance associated with topical events. I have omitted to discuss this latter aspect above, not in order to minimise its importance, but because it does not seem to me to be necessarily preponderant; there are two major risks attendant on any attempt to rely too much on current events: firstly, their dilution to the point of incoherence, since they are by definition changeable and their coverage is built on the sensational and, secondly, the negative light which is generally thrown on HR. Human rights are made to be flouted; this is partially true if we consider that they constitute an ideal to be achieved and that there they will never be fully effective in any real situation; but it is equally important for the young (and for adults) that HR should not be presented merely in terms of violations and that emphasis should be placed on the positive developments which are taking place in the very name of these rights, developments which are taking place in the very name of these rights, developments whose permanence may never be taken for granted.

Work on cases involving others must be accompanied by work on oneself; my rights are guaranteed by respect of my neighbour's rights; everyone's first right is to have one's rights respected by others. In the first part of this paper, I emphasised the importance of reciprocity, which I see as an essential element in defining the position of rights in relation to duties. HR lie between politics, which is based on the interplay of forces, and morality which is derived from ethics.

Concepts cannot be built up by the accumulation or juxtaposition of closed definitions or formulas. *They can only be formed in a slow and gradual process, by dint of contrast with their opposites.* For instance, the approach to freedom may be linked with imprisonment and prohibitions, the idea of citizenship with the idea of exclusion, the concept of equality with that of inequality and discrimination, etc. Opposites often seem more concrete owing to the situations to which they refer.

It is necessary to *analyse the distinction between knowledge and opinion.* Knowledge has rules which govern its operation and legitimacy. It has a relationship with truth and is at some remove from immediate experience, whereas an opinion carries a large emotional charge, linked to common sense and individual experience.

Finally, the whole process takes time and is beset by constant difficulties and obstacles. It is important not to be (too) discouraged when it sometimes seems that we are not making much impression. The same applies to HRE as to the defence of HR themselves, that is to say it is a never-ending struggle, in which no step forward can guarantee us against the possibility of future setbacks. This struggle is part and parcel of adherence to the system of values on which our democratic political systems agree, irrespective of differences of implementation particular to each state, and it is these values which, at all times, constitute the great wealth of HRE.

Part 2
National and
Individual Reports

2.1
Belgium
Innovation in Primary Education in Belgium – Evaluation of the School Improvement Process (A Socialization Model)

Roland Vandenberghe

2.1.1 INTRODUCTION

During the last ten years, the Centre for Educational Policy and Innovation of the University of Leuven has been involved in an evaluation project concerning the Renewed Primary School in the Dutch-speaking part of Belgium. This paper gives a short overview of the innovation project itself and of recent research activities. The research is mainly focusing on the analysis of the improvement process in the local school. Special attention is being paid to the influence of the school leader and the impact of the school culture.

2.1.2 THE RENEWED PRIMARY SCHOOL (R.P.S.)

During the school year 1972-'73, the Minister of Education decided to install a National Committee for the Renewal of the Primary School. Representatives of the three organizing bodies (the State, the Catholic Church, and important municipalities) were invited to develop a "game plan" for the innovation of the Primary School. The fact that the three organizing bodies collaborated from the start of the Project can be

149

considered as an innovation itself, as far as the development of a national innovation policy is concerned. In March 1973, the National Committee presented a policy document to the Minister in which a general charge strategy was described and some important consequences for the implementation of the change were explored.

It is important to underline the fact that for the first time in the history of educational policy in Flanders, a project was developed based on a game plan and a general strategy. This uniqueness can be illustrated by describing some of the characteristics and some structures built up in relation to the R.P.S.-project (for more information, see Vandenberghe & Depoortere, 1986).

2.1.2.1 The change involved

The R.P.S.-project is a large-scale innovation, characterized by its multidimensionality: a number of important aims must be accomplished simultaneously and coherently. Each innovation, as part of a reform, points to significant objectives. (For a detailed analysis of the characteristics of large-scale innovation projects, see Van den Berg & Vandenberghe, 1986). In other words: schools which are involved in the R.P.S.-project must cope with a bundle of innovations. The main goals of the R.P.S.-project are related to the following themes:
- enhanced integration and interdependence between the kindergarten (2,5 years - 6 years) and the elementary schools (6 - 12 years). Also enhanced continuity between the different grades of the primary school;
- increased and more effective individualization during the elementary grades particularly in relation to reading and arithmetic. It is expected that teachers adapt their teaching activities taking differences among pupils into consideration;
- enhanced contact and collaboration between classroom teachers and a remedial teacher, so that pupils with special problems in regular classrooms will be worked with more effectively. There is also an emphasis on more collaboration among teachers and pupils from different grades;
- increased emphasis on the socio-emotional and creative development of the pupils. A more child-centered approach is one of the key ideas of the R.P.S.;
- better interdependence with resources in the community environment, both in terms of the students going out into the community to learn and in terms of people from the community being used as resource-people on an ad-hoc basis within the school.

In summary: the R.P.S. represents a major reform, where the boundaries are wide, the tasks are general and amorphous, the goals are multiple and complex, and the changes required for the school as a whole are substantial.

2.1.2.2 Increasing number of participating schools

It was decided to start with only 9 schools. In these schools, teachers and classroom activities were observed and the observation data were discussed with teachers, principals and parents. These reviews have led to an indication of the main themes which should be elaborated in the "Renewed" Primary School. This process-oriented approach to educational change was unusual and unique for Belgium.

It was also decided to plan the expansion of participating schools carefully. So, between 1973 and 1980, the number expanded very rapidly (1973: 9 schools; 1976: 25 schools; 1977: 66 schools; 1980: 277 schools). An evaluation study had made clear that it was necessary to prepare schools before they entered (officially) into the R.P.S.-project. An initiation strategy was prepared by the National Committee. During the school year 1984-'85 279 schools went through a "school based self review". This was considered an initiation program for these schools, which started the project in 1985-'86. This initiation strategy has been evaluated extensive (see Vandenberghe, D'hertefeld, Wouters & Van Dooren, 1989; Vandenberghe & Vanderheyden, 1989).

2.1.2.3 A complex support structure

It was also decided to create an additional support structure, beyond the already-existing inspectorate.
At the national level a *National Steering Committee* - which has until now survived different Ministers of Education - is considered as a core structure. This National Steering Committee is made up of representatives from the major groups involved in elementary school education: the organizing bodies, the inspectorate, the parents, the unions, the universities, the teacher training colleges and the psycho-medical-centers. The Committee is responsible for the general development of the project. It is a structure linked to the decision-maker (the Minister) on the one hand, but also to the daily life in schools and classrooms on the other hand. It is not only a "thinking group", but also a group which diagnoses positive and negative developments at the school level, and which presents solutions to the Minister, who can make "official" decisions in a reliable way. Now, after more than fifteen years, it is obvious that the National Steering

Committee has had an important influence as far as the continuity of the Renewal of the Primary School is concerned.

Also at the national level, there is a team of fifteen *National External Change Agents* (5 for each organizing body). They are responsible for the national coordination and write an evaluation report every year for the National Steering Committee and for the Minister.

Besides the national team, there are also *local teams of change facilitators* (n = 96). Every team member monitors 3 to 4 schools. Mostly they organize different kinds of in-service activities, and have discussions with the principal and the staff about the way the general aims (see 2.1.) of the R.P.S.-project can be implemented in the local school. In other words, the local change facilitators try to elaborate a school-focused implementation plan.

2.1.2.4 External evaluation

The request for an external evaluation was formulated by the national team of change agents during the school year 1978-'79. The National Steering Committee established an evaluation committee which formulated several ideas and suggestions for the evaluation of the R.P.S.-project. Two evaluation teams - one at the University of Gent and another one at the University of Leuven (since the school year 1988-'89 a third team located at the University of Brussels is involved) - collected data from 1979 to 1981, which were published in 1980 and 1981. The detailed and voluminous reports have been summarized in a synthesis report. This report has been widely disseminated and discussed in some newspapers and in professional journals. The main findings have also been discussed with the members of the National Steering Committee and, during several work shops, with the national and local change facilitators.

A second evaluation study was set up from 1981 until 1985. Again several interim reports have been published and discussed. The final report was presented to the National Steering Committee in 1985. In the next sections, an overview will be given of the recent evaluation activities by the Leuven team.

2.1.2 DEVELOPMENT OF A LOCAL INNOVATION POLICY

Given the characteristics of a large-scale innovation project (such as the R.P.S.-project) one can expect that participating schools will react differently. In other words:
- it is important to emphasize that one result of the way a large-scale

152

project is organized is to provide space for reactions of a very different nature at the local school level. These different reactions are affected by the existing local context. The school has a certain size, a characteristic climate, certain procedures, and specific subgroups. To understand the change process one always needs to be cognizant of the contexts;
- the goals of the R.P.S. presented by the national authorities and developed by the National Committee, are vaguely formulated. As a consequence, schools have the opportunity to elaborate these goals in a way that is adapted to the local situation. As a result, when looking at the different schools, one can expect that the separate innovations will take on different configurations. It is not unusual to observe important differences in the innovation components between schools that are supposedly implementing the same innovations. These variations make it difficult to assess the level of implementation in an uniform way;
- given this context, one can expect that the school leader plays an important role. He or she can be considered as maker of a local innovation policy.

During the 1981-'82 school year, 101 teachers (and the principals) from 24 R.P.S.-schools were interviewed. The interview data were used to elaborate the concept of "local innovation policy" (L.I.T.). (Vandenberghe, 1987a, 1987b, 1988a). Four different types of L.I.P. were identified. Each type is briefly described below.

L.I.P. characterized by PLANNING
With the planning L.I.P., most of the efforts - of the principal as well as the teachers - are aimed at the implementation of innovations in the classroom with the purpose of improving existing practice. These efforts are coordinated by means of a plan which include specific indications of desired changes in teaching practice. The results of this L.I.P. are implementation of many changes in teaching practice in a relatively short time period.

L.I.P. characterized by INTERACTION
This L.I.P. can best be described as a process of systematic interaction. Frequent discussions and consultations are observed in these schools. These interactions occur within the school team and between the team and external change facilitators. To encourage and support discussions and to involve all school team members, the schools make intensive use of existing structures (e.g. weekly team meetings) and exchange of written information. This way of exploring the innovation(s) leads in a relatively short time to many changes in classroom practices.

L.I.P. characterized by RISK AVOIDANCE

This L.I.P. can be concisely characterized as a slow, steady and careful approach. The school team pays a lot of attention to the careful exploration of all the aspects of the innovation. They search for an answer to the problems they experience by attempting those components that everyone considers realizable in their own classrooms. Typically, there is a serious attempt to minimize the risks by involving everyone from the beginning in the search process and by keeping everyone well informed before taking concrete steps. This policy - at least in the first year of implementation - leads to only a few changes in classroom practice. The heavy emphasis on discussion and information retrieval leads in some cases to no or very minimal changes.

L.I.P. characterized by COOPTATION

With this L.I.P.-type, most concrete changes in the classroom, as well as any changes in the internal organization of the school, are primarily initiated and supported by an external facilitator. As time goes on, the school develops no collective attitude towards these changes or towards the innovation project. This kind of reaction to the innovation can lead rather quickly to many, but small changes or to few changes in classrooms. Staff members do not take responsibility for their own development. In most schools, after a while, some initiated changes disappear.

In summary:
- schools involved in a large-scale innovation project react differently in ways that can be characterized as different types of Local Innovation Policy;
- more extensive descriptions of the L.I.P.'s create a background for external change facilitators who are expected to make a first assessment of the innovation readiness of a school;
- there is a correlation between type of L.I.P. and degree of implementation: the correlation is high or average at the schools of the planning and interaction type, average or low at the risk avoidance type, and very low at the schools of the cooptation type (Vandenberghe, 1987a);
- the organizational reaction appears to be determined predominantly by the principals' intervention (Vandenberghe, 1988a). In other words, the principal is a crucial factor in the development of a local innovation policy;
- from a limited follow-up study, we know that the L.I.P. is changeable. Moreover, a detailed analysis of a L.I.P. at different moments give information about the factors which influence the nature of the L.I.P. and the related degree of implementation (Vandenberghe, 1987b).

154

2.1.4 THE PRINCIPAL AND THE LOCAL IMPROVEMENT PROCESS

One clear message of the planned organizational research is that change efforts succeed with the active support of the principal. Research in the U.S.A. as well as in Europe has underlined the importance of the role and the day-to-day activities of the principal.

One way of looking at the principal's role in improvement projects is to start from a valuable framework. During the last two years we've been developing a framework in collaboration with G.E. Hall, dean, College of Education at the University of Northern-Colorado. That framework has been used as a basis for the development of a questionnaire which has been applied in several studies.

2.1.4.1 *Conceptual framework*

Based on earlier studies, a conceptual framework was developed (Hall & Vandenberghe, 1987). Three dimensions, Concern for People, Organizational Efficiency and Strategic Sense were distinguished. For each dimension, considered as a continuum, two poles were described. In figure 1, an overview of the conceptual structure of the Change Facilitator Style Questionnaire for Principals (CFSQ) is given.

Fig. 1. Conceptual structure of the CFSQ

Concern for People	- Social-Informal - Formal-Meaningful
Organizational Efficiency	- Leave-it-to-others - Administrative efficiency
Strategic Sence	- Day-to-day - Vision and planning

What follows is a summary of a more extensive description of the dimensions.

Concern for people
The Concern for People dimension addresses the degree to which the principal emphasizes social-informal to more *formal-meaningful* interactions with the staff. At one extreme the discussions with teachers

155

deal mostly with moment to moment topics and many of the topics of interactions are unrelated to work. When work related topics are dealt with, it is done in more informal and superficial ways. At the formal-meaningful end of the dimension principal discussions have a heavy task focus and most contacts with teachers are centered around work related topics. Interventions are interconnected and the primary emphasis is on the tasks at hand. However, when there are significant personal needs these are addressed in ways that are meaningful to those that are affected.

Organizational efficiency
Accomplishing the work of the organization can be facilitated with varying degrees of emphasis on obtaining resources, increasing efficiency and consolidating/sharing responsibility and authority. Principals can try to do almost everything themselves or they can delegate most of it. System procedures, role clarity, work priorities can be made more or less clear and resources organized in ways that increase/decrease availability and effectiveness. In this dimension the principals' administrative focus is viewed on a continuum that ranges from high *administrative* efficiency by creating and making supportive procedures and systems, to *leave-it-to-others* through casual, informal and less consistent articulation of procedures and delegation of tasks.

Strategic sense
To varying degrees principals keep in mind an image of the long term view and its relationship to the monthly, weekly and daily activities of themselves and their school. Some principals are more "now" focused, while others think and act with a vivid mental image of how todays actions contribute to accomplishing long term goals, some are reflective about what they are doing and how all of their activities can add up, while others focus on the moment to moment, treating each event in isolation from its part in the grand scheme.
At the *day-to-day* end of the dimension there is little anticipation of future developments and needs or possible successes/failures. The orientation of the *vision and planning* pole is that of having a long term vision that is integrated with an understanding of how the day-to-day activities are the means that accumulate to the desired end.

2.1.4.2 The CFS-Questionnaire

A questionnaire (Likert-type) was constructed with 77 statements representing the three dimensions and the poles. After several data-analyses (factor-analysis; item-analysis) a questionnaire of 30 items has been applied in several other studies (Vandenberghe, 1988b).

Other studies have confirmed the validity of the conceptual framework and the reliability and validity of the questionnaire (Vandenberghe, Verhoelst, Staessens, D'hertefeld & Wouters, 1989).

2.1.4.3 The CFS and implementation of innovations

There are several indications that R.P.S.-schools with a high degree of implementation of innovations are monitored by principals scoring high on the formal-meaningful, administrative efficiency and vision and planning poles. A more qualitative oriented study is underway, using a semi-structured interview, in which detailed information is collected about beliefs and daily activities of these principals. This kind of information will create a basis for explaining the observations of a correlational nature.

2.1.5 PROFESSIONAL CULTURE OF THE SCHOOL AND THE LOCAL IMPROVEMENT PROCESS

Another part of the recent studies concerns the elaboration of a conceptual framework of the so-called school culture and the development of assessment instruments (questionnaire and interview).

2.1.5.1 Conceptual framework

The professional culture of a school is conceived of as "the deeper level of basic assumptions and beliefs that are shared by members of an organization, that operate unconsciously, and that define in a basic 'taken for-granted' fashion an organization's view of itself and its environment" (Schein, 1985, p. 6).
Based on an extensive analysis of the literature, we initially distinguished on a more operational level three dimensions: goal congruence, the school leader as builder and bearer of the culture, and professional relationship between teachers (Staessens & Vandenberghe, 1987a; Staessens, Vandenberghe, D'hertefelt, Van Dooren, Verhoelst & Wouters, 1989).

Data analyses have lead to the development of a fourth dimension: lack of internal support (or an internal network). The same analyses made it possible to distinguish within each dimension a number of sub-dimensions (see 5.2.).

2.1.5.2 The professional culture questionnaire

For the development of the questionnaire a number of statements were formulated representing the three initially defined dimensions. After several analyses a fourth dimension was added and several subscales were constructed.

The final questionnaire consists of 59 items. The structure is as follows:

Scale 1: The school leader as builder and bearer of the culture
   SS 1.1 The school leader as a stimulating force
   SS 1.2 Presence and articulation of vision
   SS 1.3 Clearness of vision

Scale 2: Goal-congruence
   SS 2.1 Active goal-direction of the team
   SS 2.2 Involvement in the goal-direction
   SS 2.3 Conformity

Scale 3: Professional relations among teachers
   SS 3.1 Professional interest and communication
   SS 3.2 Professional support

Scale 4: Lack of professional sustainment
   SS 4.1 Lack of professional trust in each other
   SS 4.2 Lack of a supporting structural network

Available data confirm the validity of the conceptual framework and the reliability and validity of the questionnaire (Staessens, Vandenberghe, a.o., 1989).

An analysis of qualitative date - collected in nine Primary Schools by means of an exclusive interview - is underway (for preliminary results see, Staessens, 1989; Staessens, & Vandenberghe, 1989).

2.1.5.3 The professional culture and implementation of innovations

In more recent studies we have found clear indications about the professional culture and the degree of implementation of innovations. Innovative schools are mostly scoring high on the scales 1, 2 and 3 and low on scale 4.
Nevertheless more data - quantitative as well as qualitative - are needed.

2.1.6 REFERENCES

HALL, G.E., & VANDENBERGHE, R. (1987). Change facilitator Style Questionnaire for Principals: Dimension descriptions. University of

Northern Colorado, College of Education - University of Leuven, Centre for Educational Policy and Innovation.

SCHEIN, E. (1985). Organizational culture and leadership: a dynamic view. San Francisco-Jossey-Bass.

STAESSENS,K., & VANDENBERGHE, R. (1989). De professionele cultuur in Basisscholen. Een kwalitatief onderzoek naar het intern functioneren van scholen in vernieuwing. (Professional Culture in Primary Schools: A qualitative study of the internal improvement process.) (Paper presented at the annual Educational Research Days, Leiden.)

STAESSENS, K. (1989). De schoolcultuur in basisscholen. Scholen verschillen van elkaar (The school culture in Primary Schools. Schools do differ.) Gids voor het Basisonderwijs, ORG. 1210, 5-33.

STAESSENS, K., & VANDENBERGHE, R. (1987). De cultuur van een school: omschrijving en betekenis voor onderwijsinnovatie (The professional culture of a school: definition and meaning for school improvement.) Pedagogisch Tijdschrift, 12, 341-350

STAESSENS, K., VANDENBERGHE, R., D'HERTEFELT, M., Van DOOREN, L., VERHOELST H. & WOUTERS, M. (1989). Het intern functioneren van basisscholen in vernieuwing (The internal school improvement process in primary schools). In R. Vandenberghe, & R. Van der Vegt (Eds.), Onderwijsvernieuwing. Lisse: Swets & Zeitlinger.

VAN DEN BERG, R.M., & VANDENBERGHE, R., (1986). Strategies for large-scale change in innovation: dilemmas and solutions. (ISIP-Book, 2.) Leuven; Acco.

VANDENBERGHE, R. (1987a), The Renewed Primary School in Belgium: Analysis of the local innovation policy. (Paper presented at the annual AERA-meeting, Washington D.C.)

VANDENBERGHE, R. (1987b). The Renewed Primary School in Belgium. Institutionalization of a local innovation policy: three cases. In N.B. Miles, M. Ekholm & R. Vandenberghe (Eds.), Lasting school improvement: exploring the process of institutionalization. (ISIP-Book5.) Leuven: Acco.

VANDENBERGHE (1988a). The principal as maker of a local innovation policy: linking research to practice. Journal of Research and Development in Education, 22, 1, 69-79.

VANDENBERGHE, R. (1988b). Development of a questionnaire for assessing principal change facilitator style. (Paper presented at the annual AERA-conference, New Orleans).

VANDENBERGHE, R., & DEPOORTERE, J. (1986). Het V.L.O.-project in Vlaanderen: Van plan naar realiteit. (The R.P.S.-project in Flanders: from plan to implementation.) Leuven: Acco.

VANDENBERGHE, R., & VANDERHEYDEN, L. (1989). De

schoolbetrokken analyse in het Vernieuwd Lager Onderwijs. Een onderzoek bij 'afhakende' scholen. (The School Based Review in the Renewed Primary School. A study of 'leaving' schools.) Pedagogisch Tijdschrift, 14, in print.

VANDENBERGHE, R., D'HERTEFELT, M., WOUTERS, M., & VAN DOOREN, L., (1989). Teamfunctioneren in basisscholen. (Functioning of staff in Primary Schools.) In VANDENBERGHE, R., & VAN DER VEGT, R. (Eds.), Onderwijsvernieuwing. Lisse: Swets & Zeitlinger.

VANDENBERGHE, R., VERHOELST, H., STAESSENS, K., D'HERTEFELT, M., & WOUTERS M., Begeleidingsstijl schoolleiders basisonderwijs. Ontwikkeling van een vragenlijst en validering ervan. (Change facilitator style of primary school principals. Development of a questionnaire and its validity.) (Paper presented at the annual Educational Research Days, Leiden.)

2.2
Finland
Socialization and Human Rights Education in Finland

Anniki Järvinen

This report should not be considered a comprehensive account of research and development activities in the field of socialization and human rights education. It is a selective review of information collected during a relatively brief period of time from research registers and bibliographies, and through contacts with institutions and individuals responsible for R&D in human rights education.

In this report I will present briefly, first, some projects concerned with the socialization of school children and, second, report on some experiences related to human rights education in Finland.

2.2.1 RESEARCH ON SOCIALISATION OF SCHOOL CHILDREN

In the 1970s and the 1980s Prof. Annika Takala directed a research project concerned with the development of school children's world view at the University of Joensuu. The research project included a subproject on "the formation of a geographic world view" and on "the cognitive, social and ethical development of comprehensive school children in the light of their essays".

A report on the above research project is included in a publication, edited by the Finnish Unesco Commission and coordinated by Prof. Takala, called "International Education in Twelve Countries" [1]. The publication focuses on ethical and humanistic values of international education and is part of the Unesco series of Joint Studies. - The idea of joint studies projects resulted from the Third Conference of Ministers of Education (June 1980) which adopted a recommendation requesting Member States to carry out in consultation with Unesco joint studies of a comparative nature on well-defined problem areas of common interest, preferably through their National Commissions. - Takala states in her research results, among other things, that [1] Geographical world view is built on both designative (value-free) and evaluative schemata, including ethically evaluative schemata. No actual attempts were made in the study to stimulate evaluative schemata, but part of the pupils in the seventh grade and even few fifth-graders expressed also ethical attitudes: strong sense of justice and disapproval of violations of human rights. The disapproval was directed towards events far away in time or towards countries that were geographically distant. This might be considered to imply that here the pupils have an opportunity to try out ethically evaluative attitudes with material that does not raise problems and conflicts to the extent that ethically evaluative attitudes towards the present world and towards one's own society would do. The discussion of the report focuses on the place of value-free and evaluative schemata in geography teaching.

According to research results from another research project, the developmental level of moral views is more clearly related to the level of thinking operations than to the level of concepts and the range of thinking. The girls expressing personal or operational moral points of view tend to remain at the level of immediate neighbourhood, whereas the boys with similar views tend to reach the general or the societal levels, which is perhaps indicative of the influence of sex roles in moral development. The connection between moral development and social development is further demonstrated by the fact that absence of a moral perspective and/or non-existence of moral views higher than conventional moral is most distinctly characteristic of the teasers or the bullies. [2]

Also the extensive follow-up project, conducted by Prof. Rauste-von Wright at the University of Turku, was concerned with the socialization process of young people in the 1970s. The key concept of the research project is 'world view' and the related image of man or self. At different stages of the project young people were asked about issues related to future, such as, values of life, young people's conceptions of the world when they are adults, personal future perspective and expectations regarding the world [3]. The world view has been examined in terms of both content and structure. In the latter case, a distinction was made between general and individual features, such as, how integrated the world view is, the level of consciousness, and the level of "maturity". Among research projects on socialization in progress in the 1980s, there are two that should be mentioned: Ritva Uusitalo's study "Home, school and youth culture: The possibilities of co-operation in the socialization process of young people" (University of Helsinki, Institute of Sociology, Franzenink. 13, SF - 00500 Helsinki) and Leena Alanen's study "Socialization as a concept and a problematic", which is primarily theoretical and methodological in nature. (Institute for Educational Research, University of Jyväskylä, Seminaarink. 15, SF - 40100 Jyväskylä).

2.2.2 HUMAN RIGHTS EDUCATION

2.2.2.1 Background

In Finland new legislation concerning the comprehensive and upper secondary schools came into force 1986. In these laws the goals and principles of school education have been stated in greater detail than earlier. These new laws prescribe that the school has to strive for educating its pupils so that they as individuals and as members of society are capable of cooperation and have a desire for peace. In addition, the instruction and other forms of action should be organized so that the pupils develop readiness for promoting national culture and international cooperation.

Characteristic of Finland is the great homogeneity of its population. From about 5 million people more than 90 per cent are of Finnish origin and 6 per cent speak Swedish as their mother tongue. A group of about 1,500 people speaks Lappish and has a specific culture of their own. The gypsy population is estimated to be about 6,000 (statistics not available). They speak mostly Finnish, but have maintained elements of their distinctive culture. Only very small groups belong to some other nationality or speak some other language than those mentioned.

To date the Finnish people have not had many personal contacts with members of other nationalities and language groups. There has not been much immigration to Finland. We have only a small number of political refugees and foreign workers. The situation is, however, changing all the time and both of the last mentioned groups are growing. Naturally, images of other cultures and nations are also transmitted through mass media. - In spite of generally positive attitudes, there are some signs also in Finland of hostility felt towards refugees, even though it is not an urgent issue yet.

2.2.2.2 On problems of human rights education

The Finish Unesco-Committee organized a seminar in November 1988 on Human Rights in Finland, where one work group discussed human rights education [4]. According to the working group's report, Finnish human rights education is faced with some problems, including:

- Human rights education is not a separate school subject, but is integrated into the teaching of other subjects, e.g., religion, non-denominational moral education (which is arranged for those pupils who do not participate in religious education), history, biology and geography. Therefore this subject matter may be missing from the curriculum of many municipalities.
- It may also be that even if human rights are included in the aims of a school, these aims are not necessarily realized in classroom teaching or in the human relations within the school. There are also cases of discrimination. According to the working group report, worldwide knowledge of human rights is still limited in Finland and most Finns lack facts about human rights issues.
- Furthermore, the working group states that teacher education plays a key role in the transmission of information about human rights and as an educator of teachers of children who experience discrimination. Teacher education should also import knowledge of the customs and circumstances of children coming from various cultures, and it should also encourage teachers to cooperate closely with parents of the non-native children.

2.2.2.3 Experimental and development work

In 1986 a working group of Nordic UN-Leagues and Unicef-Associations initiated a pilot experiment concerned with the teaching of human rights in the Nordic countries. In Finland the support group included also the Institute for Peace Education, the Institute for Human Rights, the National Board of General Education and the Teacher Education Department of the University of Jyväskylä. The pilot experiment involved 11 schools, a total of 2,000 pupils and a couple of hundred teachers. The schools planned independently a one-term activity program on human rights. The pilot experiment yielded a set of material [5] which is at present being translated into English. It includes the presentation of various models and work methods for carrying out human rights education.

In 1988 the National Board of General Education published a guide for international education [6], in which human rights ethics is regarded as the foundation for international education. The aim of human rights education is to make students aware of the central values of human rights, to teach unconditional respect for the value of a human being, to show how human rights are realized in different cultures, to familiarize students with the declaration of human rights and other similar agreements and to emphasize the importance of peace and development as a central human right.

In September 1989 the Finnish Unesco Commission arranged a European Seminar for ASP-teachers (ASP = Unesco Associated School Projects) 'Education for Human Rights', where the above material was used as background reading [5]. A memorandum will be prepare on the seminar, including such issues as, how Unesco, national authorities and non-governmental organizations could support the work of human rights educators, and what are the most difficult problems and obstacles in the work. The memorandum will be available in English [7] at the end of the year.

2.2.2.4 Human rights education in teacher education

Some 3,000 teachers for various sectors of education graduate yearly in Finland. According to a survey carried out by the Teacher Education Department of the University of Jyväskylä, only about 10% of teacher trainees took studies in international education in academic year 1986-87. The studies were either compulsory or optional. One third of prospective class teachers and only 2% of prospective subject teachers took part in the programme. Less than half of the study units on international education had also dealt with human rights. It is obvious that the amount of international and human rights education included in the Finnish teacher education programs is far from sufficient. Research is also needed in this area to support experimentation and development work. In autumn 1989 the Teacher Education Department of the University of Jyväskylä initiated a R&D project 'the inclusion of international education in subject teacher's training program'. [8]

In sum, development work concerned with human rights education in the Finnish school system started on a larger scale only a few years ago, and it urgently needs the support of research, which has so far been very limited in Finland.

With the support of many civic organizations, the University of Jyväskylä has submitted a proposal to Parliament for the creation of a professorship in peace and international education at the University of Jyväskylä.

2.2.3 REFERENCES

1. TAKALA, A.: Education for international understanding, co-operation and peace and educating relating to human rights and fundamental freedoms through the teaching of ethical and humanistic values, in: International Education in Twelve Countries. Publications of the Finnish National Commission for Unesco, No 37, European Joint Studies, No 9. Jyväskylä 1987, Gummerus.
2. TAKALA, A.: Peruskoulun oppilaiden kognitiivisesta, sosiaalisesta ja eettisestä kehityksestä... (On the cognitive, social and ethical development of the pupils of the comprehensive school in the light of their essays.) Joensuu 1981. University of Joensuu. Publ. of the Dep. of Education, No 16. English summary.
3. RAUSTE-VON WRIGHT, M-L.: Nuorison ihmis- ja maailmankuva X. Sosialisaatioprosessi ja maailmankuva. (The view of world and the view of human being of young people. The process of socialization.) Turku 1986. University of Turku. Publ. of Dep. of Psychology, No 36. In Finnish.
4. Human Rights in Finland - Congress in Hanasaari 8-9.12.1988. The Finnish National Commission for Unesco. English report.
5. Approaches to Human Rights Teaching. Material for schools. Helsinki 1989. Publications of the Finnish National Commission for Unesco, No 47. ISBN 951-47-2826-2. English version is in press.
6. Peruskouluopetuksen opas: Kansainvälisyyskasvatus, Kouluhallitus (A Guidebook of the comprehensive school teaching: Education for International Understanding, National Board of General Education) Helsinki, 1988. Valtion painatuskeskus.
7. European Seminar for ASP Teachers "Education for Human Rights", Kuopio, Finland 11-15 September 1989. The Finnish National Commission for Unesco. Ministry of Education, department of international affairs. English report is coming.
8. LIIKANEN, P.: Kansainvälisyyskasvatuksen sisällyttäminen aineenopettaja-koulutuksen opetussuunnitelmaan. (Inclusion of Education of International Understanding in the Curriculum of Subject Teacher Training.) Teacher Training Institution, University of Jyväskylä. Jyväskylä 1989.

2.3
Holy See
Schooling and Education for Peace

Guglielmo Malizia

The Institute of Sociology of Education at the Pontifical Salesian University has recently concluded a research on "Youth and Peace", begun in 1985 during the International Year of Youth[1]. This research was conducted among the youth of 12 European countries who were then in the final year of the upper secondary school. The results were presented at the 13th Conference of the Comparative Education Society in Europe.

2.3.1 THE HYPOTHESES AND THE SAMPLE

The research centred on these three *objectives*. The first consisted in the reconstruction of the "semantic universe" of the youth culture on peace, which refers on the one hand to concepts such as peace, non-violence, development, brotherhood, internationalism, justice, and on the other, to

---

[1]The coordinating group of the research comprised: J. Baizek, S. Chistolini, L. Macario, G. Malizia, R. Mion, C. Nanni, V. Pieroni, Z. Trenti, and was directed by G.C. Milanesi. The group, in its turn, availed itself of the collaboration of groups working in the individual countries. The general report of the research is shortly to be published by the LDC of Turin (Italy).

war, violence, underdevelopment, oppression, repression, injustice, and so on. The second objective was a census of what young people considered as "resources" for peace, that is, the strategies that in their judgement could lead to the attainment and maintenance of peace, and their attitudes towards peace education. A final objective consisted in the description of what youth considered as objective and subjective obstacles to the attainment and maintenance of peace as well as to an education for peace.

In the *general hypothesis* it had been assumed that the youth of Europe would have shown a great sensitivity to the problems concerning peace. It was further hypothesized, however, that such an attitude would have assumed the characteristics of a true "peace culture" only in a few persons who had at least partially reorganized a personal system of significance. In the research, the term culture stood for an entire system of knowledge, opinions, evaluations, disposition to action and behaviour with regard to peace and war, to violence and non-violence.

With regard to the *particular hypotheses*, we have taken into consideration only those that refer to the topic being studied in the present paper:

1 The influence exercised by scholastic and familial socialiation on involvement for peace of those interviewed would probably be scarce.
2 Among the activities that favour peace those of a cultural/formative character (discussions, debates, meetings) would be preferred to those of a political nature (demonstrations, marches, sit-ins).
3 Among the factors that could contribute to the preservation of peace the educative (rather than the political) and the social (rather than the economic) measures would be particularly stressed.

The total number of students interviewed was 13,053. They were all aged 19 years and attended the final course of the upper secondary school. Following this, the survey concentrated on a second *sample* of 4719 students, chosen at random from the first and distributed on the basis of the following variables: sex, school attended; type of studies; country of origin. The essential data of the second sample are shown in table 1a.

The sample is equally *distributed* according to sex and type of studies. The choice of the countries should have provided a complete representation of the principal existing differences. Due to financial and organizational problems it was not possible to fully satisfy such a criterion: the Scandinavian countries are not represented and the countries of East Europe are under-represented. The 12 nations that were actually selected form part of the sample with a quota of about 400 units each. The further

Tab. 1a. *Characteristics of the sample (in percentage)*

SEX		DWELLING	
Male	50.3	City	64.0
Female	49.7	Country	34.2
COUNTRY		SCHOOL ATTENDED	
Austria	8.5	State School	81.6
Belgium	7.3	Catholic School	18.4
England	8.5	TYPE OF STUDIES	
France	8.5	The Humanities	50.9
FRG	8.5	Science and Technology	49.1
Italy	8.5	RELIGION	
The Netherlands	8.5	The Catholic Church	67.5
Poland	8.5	Other Christian Churches	10.1
Portugal	8.5	Another faith / religious community	2.1
Spain	8.5	No religion	18.7
Switzerland	8.5		
Yugoslavia	8.4		

division of those interviewed on the basis of the city/country criterion differs in a remarkable manner among the countries, with the greatest urban concentration in Spain (88.5%) and the least in Austria (33.9%); however 60% of those interviewed live in areas with a population of less than 50,000 inhabitants. The state/catholic distribution roughly conforms to the percentual relationship that is verifiable in each country, with the exception of Poland and Yugoslavia wherein only state schools are to be found. The sample distinguishes itself also for the inclusion of a large number of students who declare themselves as belonging to the Catholic church. In fact, the countries in which the research was conducted are characterized by the majority of Catholics among the inhabitants or by the presence of a significant Catholic minority.

The *social origin* of the students is prevalently middle or upper class. The majority of the fathers and approximately 40% of the mothers hold an upper secondary or a university degree. The sample is outstanding also for a considerable diffusion of dependants both among the fathers (75.8%) as also among the mothers (53.9%) and for the rather high percentage of mothers who work outside the home. The distribution according to the professional sectors, that reveals a concentration of fathers in the secondary, tertiary and advanced tertiary sectors, also confirms the fact that the majority of the sample come from the middle classes.

The sample *neither includes* all the European youth no all the youth of the nations interviewed. The cohort of age is restricted since only 19 year old students who were then attending the final course of the upper secondary

Tab. 1b. *Distribution of the sample according to: the parents' educational qualification, the parents' employment and the parents' type of employment (in percentage)*

	FATHER	MOTHER
EDUCATIONAL QUALIFICATION		
None	14.0	17.0
Completed compulsory schooling	32.1	37.5
Completed upper secondary school	24.3	24.8
Qualified at university level	23.4	12.2
EMPLOYMENT		
Director, executive, professional	15.1	2.3
Teacher, military officer	6.5	8.4
High grade employee	12.3	6.0
Middle grade employee	12.3	12.5
Technician	12.9	4.9
Ordinary worker	7.6	4.1
In service (doorman, domestic cleaner)	0.5	4.5
Other employed persons	5.6	11.2
Owner, professional, manager	7.9	1.5
Artist or craftsman	2.0	0.5
Skilled technician	0.8	0.2
In business	3.0	2.3
Selfemployed (horticulture or agriculture)	3.5	2.4
Other selfemployed	1.5	1.8
Not employed (retired, pensioner, sick, at home)	5.9	30.2
SECTOR OF EMPLOYMENT		
Agriculture, fishing, mining	8.4	5.4
Industry and crafts	26.5	9.0
Commercial, transport, tourism	16.1	11.7
Public service or administration	27.3	24.3
Advanced technology (electronics, computers)	5.4	1.0

school are included; besides, working youth, those unemployed or those in search of a first employment have been excluded. We have already referred to the reason for the limited number of nations represented in the research. It must be added that the representation in the first sample was not very satisfactory.

Notwithstanding the aforementioned shortcomings, the sample is sufficiently *adequate* with regard to the criteria such as sex, school attended, the type of studies, the distribution of those interviewed across the country and the religion. In short, it reflects in a satisfactory manner a cohort of European students, predominantly belonging to the Western geographical region, of a middle or high social origin, in the age-group situated between the end of adolescence and the onset of youth. Within such a frame of reference, a prudent extension of data to the whole corresponding group, beyond the sample, is possible.

2.3.2 A RATHER INADEQUATE EDUCATION

One of the objectives of the research was to verify the influence that the principal *agents of socialization* exercise on the formation of a peace culture among youth. In particular, traditional institutions like the family, the school and the associations, and the more recent forms of aggregation such as groups and movements were taken into consideration. The results are more encouraging than the predictions contained in the hypotheses, although the inadequacies are many and serious.

Those interviewed were asked whether the education received at school had stimulated them to interest for peace. The students of the sample were of a *divided* opinion: 45.5% responded positively, while 52% bordered on the negative (cfr. tab. 2). The percentage of those favourable proved to be greater than that hypothesized: the influence exercised by scholastic socialization on involvement of those interviewed for peace had been assumed as being minimal. The upper secondary school contributes to the formation of a sensitivity and an awareness of the problems of peace in a consistent minority of students that comes close to 50%. It must be added that on the basis of more sophisticated statistical analysis this same group manifests a rather high ranking on the scales of religion, of peace culture, of non-violence, of idealism and of involvement for peace: in other words, there appears to exist correlation between a mature peace culture and a school education that is efficacious on the level of values, even though it is difficult to establish with precision the relative impact of the latter.

However we cannot ignore 52% of those interviewed than offer a *negative* evaluation concerning the stimulating function of the school in relation to education for peace. At any rate, it is true that the positive influence of the family does not seem to be superior to that of the school: in less than one third of the families the problems of peace and war are often dealt with and yet not even one fifth of the family members participate in activities and initiatives in favour of peace. Besides, the incidence of the socialization carried out in associations, even if higher, seems far from being fully satisfactory.

If one studies the results of the question from the point of view of the *national* samples, one notices three distinct groupings. The students who hold that the school education had stimulated in them interest for peace are an absolute majority in the Federal Republic of Germany, Austria and Spain and in the FRG come close to 60%. Switzerland, France, England and Poland are placed slightly above the average score: Belgium coincides with the average and Italy is a a little below the average. Two thirds of the sample in Portugal and Yugoslavia and 60% in the Netherlands declare that the education in school has not contributed to the arousal of attention or of sensitivity to the subject of peace.

The *Catholic school* has a favourable influence on education for peace. A consistent absolute majority of its pupils - 60.7% - believe that the formation received therein had made a positive contribution, while the percentage comes down to 42.1% in the state schools. The schools that are situated in less urbanized areas also present similar characteristics. The religion to which one belongs also has an impact in the sense that the members of the Catholic church and of the other Christian churches express more favourable opinions than do the students of other religions or those areligious. Moreover, the opinions become more critical as one ascends the social scale. The sex and the type of studies do not however seem to exercise a significant influence.

Among the *initiatives* that had stimulated interest for peace, discussions, debates and meetings organized by the school themselves, occupy the first place: these types of activities are indicated by more than 80% of the respondents. The same initiatives receive less emphasis in Yugoslavia - indicated only by an absolute majority - and in Poland - only by two thirds. Less than one third of the total states that their own school favours participation in discussions, at debates and at meetings outside the school. The percentage rises approximately to 40% in Yugoslavia and in Poland and is more than one third in England, Portugal and France; on the negative side, we find the Netherlands and Belgium. Only 12.9% affirm that the school education had encouraged effective participation at demonstrations, marches and sit-ins. Yugoslavia is a striking exception - 55.5%; percentages higher than the average, even though less revealing, are offered in Poland, Belgium and Italy.

The formative activity of the school in the realm of peace seems *limited* to making the students aware of the problem. There are very few occasions in which the education provided had led to a real militancy outside the school. The datum however conforms to the general orientations of the research data that reveal a sensitivity to peace that is more "vocal" than "action oriented".

2.3.3 THE SCHOOL IS NOT A PLACE OF "SYMBOLIC VIOLENCE"

Those interviewed were asked to indicate which among eleven words (struggle, family, power, religion, competition, school, aggression, state, conflict, church, sport) appeared to be closer/further with reference to the concept of *violence*. The number "1" indicated the position that was closest, while "5" the furthest and "2, 3, 4" were the intermediate positions. The terms that were considered as being nearest to the idea of violence were aggression (X = 1.46), struggle (1.75), conflict (1.81) and power (1.89) (cfr. tab. 4). The replies highlight two dimensions, one more individual and psychological, and the other more structural and social.

The others were placed at varying distances in the following order: the family (4.34), the church (3.87), sports (3.76), the school (3.71), the religion (3.60). A great majority of the respondents refused the pairing of violence and the traditional and new agencies of socialization. Of particular interest is their attitude to the educational system. During the period of contestation it was commonly believed that the school was a privileged place for "*symbolic violence*": the youth of the sample on the contrary reveal that they have a positive image of the school. This tendency is predominantly manifested among the students of Spain, Italy, the Netherlands, France and Portugal; it is interesting to note that among these nations are to be found those in which the 1968 contestation was most actively felt.

A confirmation of this trend comes from other questions of a similar nature. The youth were asked to select from a list of determinate words those which were closest to or further from the concept *"peace"*. In this case the term "school" was excluded and the word "education" was introduced. Non-violence (1.48), freedom (1.64), reconciliation (1.63), tolerance (1.78), and equality (1.79) were the words that came close to the concept of "peace". Instead, the more distant words were colonialism (4.35) and capitalism (3.87).

On the intermediate positions of closeness to the concept we find words such as democracy (2.06), education (2.34), internationalism (2.56), religion (2.84), development (2.85), ecumenical movements (2.86) and progress (2.89). The position of education on the scale indicates that the concept has attained a positive consideration with regard to peace among the majority of the youth, even though not among all of them. The countries that register a greater progress in this direction are placed in the following order: Spain (1.86), Italy (1.88), Poland (1.89), Portugal (1.96). The data partly confirm certain tendencies noted in the previous question.

2.3.4 THE SCHOOL: STRATEGY FOR PEACE

Among the objectives of the research, the opinion of the youth was sought as regards *strategies for peace*, both on the structural and personal level, through a realistic confrontation with the factors that threatened it. The questions that dealt with the topic were articulated on three levels, world, national and interpersonal, and on each dimension an evaluation to its validity and its feasibility was called for.

The introducing of *education for peace* in the school is considered a valid strategy for the maintaining of peace among the different social groups of a nation by three fourths of the sample, and more than four fifths believe it to be feasible (cfr. tab. 5). It is a strategy that represents the greatest degree of convergence between validity and practicability. If it is true that school education has encouraged only one half of the youth to interest for peace, the majority believe that a reform of the curriculum, which tends to give education for peace an adequate and efficacious role, could contribute in a significant manner to the maintaining and the development of peace. Undoubtedly, those interviewed do not seem to have considered the methodological problems associated with the introducing of education for peace in the school, that is, how to carry out the planning of the curriculum at various levels: this however is a justifiable underestimation given the fact that the youth are non involved in its planning.

The Spanish, Portuguese, Italian, and to a lesser extent, the Yugoslav sub-sample, emphasize the feasibility of the strategy: these same countries - with the substitution of Yugoslavia for Austria - declare themselves as being more convinced of its practicability than the average of the total sample. On the negative side of the scale one notices that approximately two thirds of the Dutch youth are of the opinion that the strategy is not valid and a great majority of the English do not consider it feasible.

Three fourths of the respondents agree in affirming that the *promotion of education* positively contributes to the establishing of peaceful relationships between individuals. The research confirms a trend already noted above: the antipedagogism of the student movements during the period of contestation, their total distrust of the school and the possibility of the reform seem to be on the decline. The youth express an almost total trust in the school's ability to renew itself: 84.6% state that the proposal is practicable. The greatest supporters of the belief that promotion of education as a strategy to maintain peaceful relations among individuals is valid and feasible are the youth of Italy, Portugal and Yugoslavia. The Dutch and - to a lesser extent - the Spaniards are more critical on both

174

fronts. Even the Austrians and the Germans are less convinced of the positive value of the proposal.

2.3.5 CONCLUDING REMARKS

Although it has not been possible to provide all the relevant data, it seems opportune to point out in the first place that the research has *confirmed* the general hypothesis. A satisfactory level of attention and sensibility to the problem of peace has been verified among the students. At the same time, difficulties that youth encounter in the passage from a stage of awareness, sometimes acute, of the problems of peace to one of a mature culture, have come to light: besides, a remarkable gap has been noticed between the capacity for reflection and evaluation and the level of action, that is, the realization of the proclaimed ideals in daily life. This does not exclude the presence of smaller groups that have developed a more profound, consistent and committed awareness of peace.

The education received at school made an *important* contribution to the awaking of sensitivity to problems of peace in nearly half the members of the sample. The school exercises this positive influence especially through debates, discussions, meetings and much less through active involvement. Even though the data have provided a rather positive image than that assumed in the particular hypotheses, we cannot forget that more than half the youth presented a totally negative evaluation as regards education to peace imparted in the school.

The majority of students interviewed reveal a *trust* in the potentiality of the school system. The introducing of education to peace in the school and the promotion of instruction are considered valid and practicable strategies for the maintaining and the development of peace in society and between individuals. In short, the school in itself does not seem any longer to be considered either a place of "symbolic violence", or an irreformable structure, or an instrument of social reproduction.

The data of the national sub-samples moreover do not reveal a reversal with respect to the general orientations: the differences are of *degree* more than of direction. One may however recall a consistent absolute majority of Germans who positively evaluate the education received in school, the equilibrium between reflection and action in the initiatives of the school towards peace revealed in Yugoslavia, and the emphasis in Spain, Portugal and Italy on the school's potentiality of building a future of peace.

2.3.6 BIBLIOGRAPHY

ARON R., Paix et guerre entre les nations, Paris, Calman-Levy, 1984.

BOBBIO N., Il problema della guerra e le vie della pace, Bologna, Il Mulino, 1979.

BOUTHOUL G., La pace tra storia e utopia, Roma, Armando, 1976.

CALVARUSO C.-S. ABBRUZZESE, Indagine sui valori in Italia, Torino, SEI, 1985.

GALIZZI G., (Ed.), Lo sviluppo dei popoli e il nuovo nome della pace. Milano, Angeli, 1984.

GALTUNG J., Ci sono alternative!, Torino, Gruppo Abele, 1986.

GROI U., Natura e orientamenti delle ricerche sulla pace, Milano, Angeli, 1979.

IMPEDOVO G., (Ed.), Educazione alla pace, Roma, Citta Nuova, 1981.

MILANESI G., (Ed.), Educazione alla pace, Roma, SEI, 1985.

IDEM, I giovani e la pace, Roma, LAS, 1986.

MION, R., Per un futuro di pace, Roma, LAS, 1986.

2.4
Italy
Education for Human
Rights and Democracy

Ministry of Education

Interest is growing in Italy concerning the problems of education about democratic values and human rights both within and outside the framework of formal schooling.

The current stage may be described as one of discovery and growing awareness, not (yet) having reached the stage of actual achievement of adequate programmes and methods.

The Ministry of Education is now committed to a continuous effort to introduce a "human rights" element into civic courses at both primary and secondary schools.

Non-governmental organisations (eg. Amnesty International, Lega per l'ambiente, (World Association for) the School as an Instrument of Peace, Mani Tese, and development co-operation associations) have shown a great deal of interest and initiative in this field. The involvement of religious institutions is particularly notable: in most Italian dioceses there have been for the last two or three years specialised centres and courses for the socio-political training of young people. The content of these programmes concentrates chiefly on the themes of human rights, peace, democracy, the environment and European integration.

Local and regional bodies are also beginning to become involved in the field of human rights education. Among the municipal authorities we should mention the exemplary initiative of the those in Boves (Piedmont) and Castel di Codego (Veneto) which have created actual "peace schools" and the initiative of the Province of Padua which has for the last two years, with the co-operation of the University, been organising an annual course (30 seminars) on the theme of "The Experience of Democracy" for secondary schools (300 pupils).

In many local areas, formal joint ventures are being set up between local authorities, non-governmental associations and schools (secondary and primary) with the aim of offering human rights and peace studies courses.

It should be pointed out that, at central government level, the Ministry of Education concentrates on the study of human rights *as a branch of civics*, whereas local institutions and associations concentrate *directly* on human rights, democracy, the environment, peace and development; their courses are entitled "education for peace and human rights" or "Human rights education".

It may be noted that at the level of local institutions and associations, the approach taken by UNESCO in the 1974 Recommendation is the most widespread. Certain regions of Italy - in particular, Friuli-Venezia Giulia, Veneto, and the autonomous provinces of Trento and Bozen - have recently enacted specific legislation "for the promotion of culture and peace" (founded on respect for human rights).

The most important and the most complete of these is the act of 30 March 1988, No 18, of the Veneto Region. In Section 1, paragraphs 2 and 3, it is laid down that the Region "shall promote culture and peace by means of cultural and research initiatives, education, co-operation and information aimed at making Veneto into a land of peace. To that end, the Region shall take decisions on independent initiatives and shall promote those of local institutions, associations, cultural institutions and charitable and international co-operation groups working in the Region". Article 3, entitled "Initiatives in the research field" lays down that "1. The region shall encourage research concerning: a. peace, human rights and the rights of peoples; b. new relations between industry, scientific research and technological innovation with a view to the development of culture and peace; c. education and teaching aimed at the production of peace studies courses for schools. (...) 3. The region shall encourage the distribution in schools of the results of such research and of the teaching materials thus produced".

The special committee set up to implement this regional Act comprises, inter alia, representatives from the three universities of the Veneto region (Padua, Verona, Venice).

Italian universities have only recently started to take an interest in human rights studies, taking a lead from the increased awareness and initiative on the part of non-governmental organisations and grass-roots movements.

There is at present in Italy only one structure which specialises in this field: the Study and Training Centre on Human Rights and the Rights of Peoples, at the University of Padua. This centre organises, inter alia, annual training courses aimed principally at teachers and NGO officials.

On the initiative of this centre, a specialist School for the study of "institutions and techniques for the protection of human rights" (course length: 3 years) was set up in 1988. Graduates ("laureati") of any university faculty may be admitted. One aim of the School is to train teachers and school planning experts in the specific field of human rights.

A group of university teachers has already taken part in the School's course, resulting in their becoming involved in research programmes on the various aspects of human rights and in particular on education about human rights, democratic values and peace.

One of these research programmes, co-ordinated by the University of Padua centre, is entitled "Peace and human rights culture: teaching and research", involving teachers and researchers from twenty universities. The National Research Council (CNR) has expressed its interest in this initiative and will be sponsoring it.

It must be pointed out that university researchers in this field are generally interested in legal rules, institutions, and, more recently, policies. Most of the research currently being carried out has arisen from initiatives by lawyers, philosophers, historians and also a few political scientists.

For their part, educationalists seem to be chiefly interested in the improvement and "aggiornamento" (updating) of civics courses, looking from a viewpoint which is still excessively institutional (training the citizen rather than the human being as such).

At present, with the exception of a few small groups of educationalists who are aiming to develop specialised "human rights" lines of research (eg. in the Universities of Padua, Bari, Milan), the more advanced current of Italian education theory are concerned with themes such as:

- education for democracy in politics (teaching to "think politically");
- education for peace and international understanding;
- education for life in a community, in other words teaching mutual comprehension of the problems of immigrants, minorities, etc.;
- education for development;
- environment education.

One very interesting line of research is the one known as "inter-cultural studies" (University of Rome, Frascati Centre).
These various lines of research each take into account one or other particular human right, but do not relate to "human rights" as an overall guiding principle for study.

There is also a new type of study entitled "education for world citizenship", taught mostly outside the framework of formal schooling which seems to be increasingly in line with the guiding principle of "human rights and the rights of peoples".

It is important to point out that most of the teachers and university researchers working in this field are involved outside the confines of formal education as active members of associations and movements. Very often their work in this field spills over into the world of universities and educational establishments.

The Study and Training Centre on Human Rights and the Rights of Peoples, and the specialist School for the Study of Institutions and Techniques for the Protection of Human Rights, at the University of Padua, are obviously working within the university world, but maintain close links with associations and institutions at local, district and regional level. The aim of this university structure is to encourage the various currents of research (peace, democracy, environment, development, etc.) listed above, to move towards the guiding principle of the "rights of man and of peoples".

The Centre's approach, which is at the basis of the national research scheme involving 20 universities, concentrates on the following points:

- the international dimension of the content of human rights education (the

age of planetary interdependence and the internationalisation of human rights);
- the inter-disciplinary dimension of human rights research and education;
- the restoration of freedom and sovereignty to human beings and communities at all levels "from the village to the UNO";
- the guaranteeing of the rights of human beings and of peoples as an essential aim and source of legitimacy, for any institution or system (national or international);
- the indivisibility and inter-dependence of all human rights;
- human rights education as training for action, ie. for the performance of the specific tasks required by democracy, service, solidarity and peace;
- the indivisibility and inter-dependence of internal and external roles, in other words the principle that human - both individual and collective - represents a continuum throughout all levels from the village to the UNO;
- the practice of transnational relations recognising the need of individuals and peoples to express themselves in a way different from that of state and inter-state institutions;
- the cultivation of a new humanism, a concept for the future aimed at guiding and transforming rules and institutions on the basis of the principle of human rights;
- a strategy for "pan humanisation" of systems and institutions, the construction of a new democratic international order;
- the practice of international democracy: in other words the achievement of sovereignty for individuals and peoples as such - not merely as citizens of their countries and constituent elements of states - by means of forms of popular participation in the functioning of inter-governmental organisations.

2.5
United States of America Human Rights Education: Alternative Conceptions

Norma Tarrow

2.5.1 SUMMARY

Cognizant of *alternate conceptions* of human rights and human rights education, this chapter attempts to:
- Provide a working definition of human rights education and its components.
- Offer an overview of various rationales for its inclusion in the curriculum.
- Discuss alternative types and levels of programmes.
- Relate human rights education to other current social education movements.
- Point out problems in implementing human rights education programmes in the schools.

Human rights are often in the news - more in regard to their violation than their observance. In addition to the issues raised by *glasnost* in the Soviet Union, by terrorists in Iran, Ireland and Spain, by political upheaval in Latin America, famine in Africa, and refugees and migrants all over the world, the fortieth anniversary of the *Universal Declaration of Human*

Rights (1988), the thirtieth anniversary of the *Declaration of the Rights of the Child* (1989), and the birth of the *International Convention on the Rights of the Child* (1990), have produced a burgeoning interest in the international protection of human rights. Recognition that an educated public is the greatest guarantee of human rights and, in effect, serves as its ultimate sanction (Humphrey, 1987: Ray and Tarrow, 1987) has fostered interest and efforts in the field of human rights education, on the part of international, national as well as state, provincial or county officials.

2.5.2 WHAT IS HUMAN RIGHTS EDUCATION? (AND WHAT ARE HUMAN RIGHTS?)

Human rights education has been defined as the conscious effort, both through specific content as well as process, to develop in students an awareness of their rights (and responsibilities), to sensitize them to the rights of others, and to encourage responsible action to secure the rights of all (Tarrow, 1988, p. 1).

This deceptively simple definition opens a Pandora's Box of alternative conceptualizations. First, and foremost, one must determine what constitute human rights.

One expert has offered the following definition: Human rights are "those entitlements which are basic to being human and are not connected to the accident of being born in a certain country or with skin of a particular colour", (Torney-Purta, 1988, p. 16). She points out that there is a basic core of universally agreed-upon human rights as well as a collection of documents in which they are expressed. *Basic rights* deal with the dignity and worth of the person; *civil and political rights* with the right to participate in self-government; and *social, economic and cultural rights* with such matters as the right to work, to maintain one's culture and language, and to receive an adequate education. Obviously the right to education is pivotal since it is essential in socialising the younger generation into understanding their rights, relationships to others, and responsibilities in terms of safeguarding human rights.

Different societies, however, define rights in terms of their own historical experience, their value systems, and the political and economic realities of the current era. The history of human rights has continuously been ruffled not only by the differing perspectives of eastern and western nations as well as developing and industrialized societies, but also by the different emphasis on "individual" as opposed to "group" rights inherent in these different societies. (Ray and Tarrow, 1987)[1]. And, in the succinct words of

184

one author, "The common denominator of human rights is that every government claims to honour human rights, while postulating that political systems different from their own do not" (Obermeier, K. 1986, p. 113). Thus, it is equally obvious that the form and content of human rights education will depend on the values into which a particular society wishes to socialise its children.

As for the documents referred to by Torney-Purta, there are those who challenge some of the critical international agreements on human rights on the basis that they reflect a particular (western) cultural perspective. In response others challenge these viewpoints as a type of skeptical relativism that can be used to justify violations and advocate wholesale rejection of the human rights movement. (Coomaraswamy, 1982). The author subscribes to the view of renowned legal and human rights expert, John Humphrey, that the Universal Declaration of Human Rights meets the criteria for being considered part of the customary law of nations, and that, as such, it is binding on all states, regardless of whether or not they had a voice in its adoption (Humphrey, 1984). The perspective of this chapter derives from the author's experiences in working with human rights educators in the U.S., Canada, Spain and the Council of Europe - the latter of which has taken the lead in defining the field, disseminating materials and developing a cadre of committed professionals implementing programs in their own countries.

The definition offered at the beginning of this section also referred to *content* and *process*. Clearly these may also be open to varied conceptualizations. Within the school setting, in its broadest sense, "content" refers to everything children learn about a subject. "Process" refers to the means and methods (both direct and indirect) by which this learning is accomplished. The United States National Council for the Social Studies curriculum guidelines offer a simple, yet complete categorization of the content of social studies curriculum, which appears to have wide applicability and acceptability. The four elements included are *knowledge*, (elsewhere used synonymously with 'content' in a more narrow interpretation) *skills, values and social participation* (NCSS, 1979).

---

[1] See also KARTASHKIN, C. The socialist countries and human rights, in VASAK, K. (Ed.) *The International Dimensions of Human Rights* Vol. 2, 1982, Westport, CT, Greenwood Press; MOWER, A., Jr. Human Rights in the Soviet Union, in: NELSON, J.L. and GREEN, V.M. (Eds.) *International Human Rights: Contemporary Issues*, Stanfordville, NY, Human Rights Publishing Group, 1980, pp. 199-227; ORWIN, C. and PANGLE, T. The philosophical foundation of human rights, in: PLATTNER, M.F. (Ed.) *Human Rights in Our Time*. Boulder, CO: Westview, 1984, pp. 1-22; and SEYMANSKI, A. *(Human Rights in the Soviet Union*, London: Zed., 1984.

According to the Deputy Director of Education, Culture and Sport of the Council of Europe, the common core of *knowledge* of human rights education should include[2]:

- The main categories of human rights, duties, obligations and responsibilities. (The idea of rights should be matched with that of responsibilities to others, to the community and to humanity as a whole).
- The main international declarations and conventions on human rights (e.g. the *Universal Declaration of Human Rights*, the *Declaration (and Convention on the Rights of the Child*, the *European Convention for the Protection of Human Rights and Fundamental Freedoms*, the *American Declaration of the Rights and Duties of Man*, the *Universal Islamic Declaration of Human Rights* and the *Banjul (Africa) Charter of Human and Peoples' Rights)*.
- People, movements and key events in the historical and continuing struggle for human rights (e.g. Ghandi, King, Mandela; civil rights movements, women's movements);
- The various forms of injustice, inequality and discrimination, (e.g. racism, sexism, terrorism, and genocide) (Stobart, 1987). The *skills* developed by human rights education are similar to those of any good social studies programme, and include:
- Communication (written and oral expression, discussion and listening) and using the tools of the discipline (e.g. reading and interpreting maps, charts, graphs, reference material, etc.)
- Critical thinking (collecting and analyzing material from multiple sources, identifying different perspectives, distinguishing between fact and opinion, detecting prejudice and bias, reaching logical, fair, and balanced conclusions)
- Social skills co-operation, conflict resolution, forming positive relationships).

The fortieth Council of Europe Teachers' Seminar, pointing out that knowledge goals are inappropriate for younger children, identified certain *values* that are appropriate at every level but that form the major focus of human rights education in early childhood. These include:

- Development of positive self-image.
- Increased awareness of one's own environment.
- Openness towards others.
- Acceptance of (and respect for) differences.
- Recognition of fundamental similarities. (Abdallah-Pretceille, M. (1989)

And lastly, the element of *social participation* brings us back to the NCSS

[2] The author has taken the liberty of modifying the examples offered by STOBART within the parentheses.

Guidelines:Knowledge without action is impotent. Whatever students of the social studies learn should impel them to apply their knowledge, abilities, and commitments toward the improvement of the human condition. A commitment to democratic participation suggests that the school abandon futile efforts to insulate pupils from social reality, and instead, find ways to involve them as active citizens. (NCSS 1979).

We have now dealt with human rights education in terms of *content* in primary and secondary schools. Starkey (1984) views this as one aspect of a tripartite structure, pointing out that human rights education also pervades both *extracurricular activities*, as well as, what he refers to as the 'hidden curriculum' - the *ethos and organization of the school*. This brings us to the matter of *process*. The messages about human rights that children receive from the way teachers and administrators communicate and interact with them, with each other, with staff, and with parents; the extent to which students are allowed to participate in decisions affecting them; the emphasis on individualization and co-operation rather than competition; the ethnic makeup of the school; the provision of relevant, integrated, interdisciplinary learning activities calling for problem solving and active involvement - all these processes are at least as significant as the content of the curriculum. A human rights curriculum requires an environment based on human rights principles.

Such an educational environment respects human dignity, similarities and differences, and group, national and universal values. Such teaching-learning strategies stress self-discipline, active and co-operative learning, as well as critical thinking and conflict resolution. Torney-Purta (1982) points out that there is considerable research support for the value of an open classroom climate and student participation, and cooperative learning; that "proclaiming" human rights in authoritarian classrooms conveys a contradictory (and hypocritical) message. The Council of Europe's *Learning for Life* programme recommends that schools have a council with elected representatives, even in primary schools, to make recommendations on matters affecting pupils. It calls for clear policies on racism and sexism; participation of teachers, parents and school employees in school affairs; and suggests that pupils should be able to meet in their spare time and express themselves orally, through the arts and sports, and in school newspapers (*Learning for Life*, 1984).

Human rights education in primary grades in well suited to the current emphases ... on development of positive self concept, democratic values (working in groups, sharing, taking turns, respecting the rights of others, co-operating in solving problems), basic civic values (fair play, good sportsmanship, respect for the opinions of others) and cultural diversity. Human rights education in middle and upper grades capitalises on current interest in educational strategies for interactive learning including group processes, co-operative learning, conflict resolution, and techniques such as role playing, simulation, etc. ... (while) middle and upper grade content expands upon themes in current curriculum guides, including civil rights, international agreements, stereotyping, and examples of human rights violations in the past and present (Tarrow, 1988, p. 8).

With this clarification of terminology, perhaps the original "simple" definition of human rights education as 'the conscious effort, both through specific content as well as process, to develop in students an awareness of their rights (and responsibilities), to sensitize them to the rights of others, and to encourage responsible action to secure the rights of all' (Tarrow, 1988, p. 1) can be universally acceptable.

2.5.3 WHY HUMAN RIGHTS EDUCATION?

"... That every individual and every organ of society, keeping this Declaration constantly in mind, shall strive by teaching and education to promote respect for these rights and freedoms ..."
Preamble, Universal Declaration of Human Rights

Education shall be directed to the full development of the human personality and to the strengthening of respect for human rights and fundamental freedoms. It shall promote understanding, tolerance and friendship among all nations, racial or religious groups, and shall further the activities of the United Nations for the maintenance of peace.
Article 26(2), Universal Declaration of Human Rights

Based on recognition of the Universal Declaration of Human Rights as part of the customary law of nations, these two universally accepted statements should serve as the rationale for the teaching of human rights in all countries. However, in response to the mandate of Article 26, and as authorised by its own Constitution, the 18th session of UNESCO adopted an important policy statement in 1974 - revolutionary in accommodating to the multiple perspectives and conceptualizations of its representatives and member States. The 1974 UNESCO *Recommendation concerning*

[3.] Note that UNESCO Recommendations are passed by majority vote of the General Conference, coming into force as soon as adopted. Differing from UNESCO Conventions, (See Note 5) a Recommendation is a non-obligatory statement of principles or norms that member States are *invited* (but not legally compelled) to implement.

Education for International Cooperation and Peace and Education relating to Human Rights and Fundamental Freedoms" not only affirms the link that exists between international education and human rights education; it also calls on governments, education officials and teachers to recognize this interrelationship and to translate it into action through appropriate educational policies, practices and programmes" (Buergenthal and Torney, 1976, p. 2).

Intended as a major guideline of educational policy, its enumeration of objectives is provided in §4 and consequent implications for curriculum planning in §5 and §18. The credibility of this document as the major rationale for human rights education in all countries is supported by the fact that:

Member States are 'required by the UNESCO *Constitution* to bring the recommendation to the attention of those national agencies in their country that are empowered to regulate and act upon the subjects dealt with in the recommendation. The UNESCO *Constitution* also requires the Member States to file periodic reports with the Organization detailing what action, if any, they have taken to give effect to the recommendation (Buergenthal and Torney, 1976, p. 4). It is significant to note that the definition of 'education' (provided in §1 (a) is extremely broad and that formal schooling is not the only type of education included. The document continuously stresses the relationship between human rights education and global education (§4), civic and moral education (§10-16) peace education (§6), and intercultural education. — through the study of domestic ethnic (as well as foreign) cultures (§4b, §17, §21, §22, §33). From the standpoint of process, and innovative, active and interdisciplinary approach is stressed (§20). The document makes a strong case for beginning human rights education in preschool and continuing through secondary and adult education (§24), for pre and in-service teacher training (§33-37), for development of educational materials (§38-40) for research (§41-42), and for international co-operation (§41-45).

At the regional level, citing Article 3 of the *Statute of the Council of Europe*, of 5 May, 1949, Eide points out that respect for human rights is not only a proclamation of guiding values but a condition of membership in the Council of Europe (Eide, 1989). The Council prioritized its project in human rights education on the basis of *Resolution (78)21 on the Teaching of Human Rights*, adopted by the Committee of Ministers (the highest political body of the Council of Europe) on 25 October 1978. As a result the Council of Europe's Council for Cultural Co-operation (CDCC) embarked on a five year project, working with teachers, specialists and non-governmental organizations — culminating in the Vienna Symposium on 'Human Rights Education in Schools in Western Europe'.[4]

In 1985, the Committee of Ministers made a point of reaffirming the principles of their 1978 Resolution and of formally adopting most of the agenda of the Vienna resolutions in *Recommendation No. R (85) 7 on Teaching and Learning about Human Rights in Schools*, this time specifically delineating skills (Section 2) and knowledge (Section 3) to be included in the content as well as recommended processes (Section 4) and implications for teacher training (Section 5) in the field of human rights education.

Every year since 1976, the Intergovernmental Programme of Activities of the Council of Europe has included a special section on human rights education. Since 1977, the Committee of experts for the promotion of education and information in the field of Human Rights (DH-ED) has coordinated conferences, teacher training workshops and development of materials. The third medium term plan (1987-1991) of the Intergovernmental Programme includes human rights education policy. The philosophy of the Council of Europe is that the promotion of human rights education and information should be *unremittingly pursued* (Committee of experts, 17 June 1987).

Beyond the Council of Europe, this author's experience in the United States and Canada permits a description of the rationale for human rights education in both of these western democracies. Canada's 1982 *Constitution* and *Canadian Charter of Rights and Freedoms* entrenched individual and group rights provided for by the 1960 *Bill of Rights*, the 1969 *Official Languages Act* and the 1977 *Canadian Human Rights Act*. At the provincial level, all Canadian provinces have had human rights legislation in place since 1978 (Crombie, 1988). Human Rights Commissions not only play an important role in redress of grievances but focus heavily on the educational arena. The rationale for Canada's strong support of human rights education is offered by the former Director of the Canadian Human Rights Foundation. He points out that:
- Canada is signatory to the *Universal Declaration of Human Rights* as well as the two international Covenants (which constitute the

---

[4] Recommendation 1 of the Vienna Symposium contains detailed suggestions for content (Section 3) and process (Section 4) of human rights curriculum as well as guidelines for the initial and in-service training of teachers (Section 5). Attention is given to the development of materials (Section 5.3) and methods, to evaluation (Section 5.4) and to sensibilities regarding education in the realm of political activity. (Section 3.5) Recommendation 2 suggests further action to be undertaken by the Council of Europe (See STARKEY, 1984).

International Bill of Rights) — thus indicating an international legal obligation.[5]
- Human rights are not an amorphous set of principles. They are law - with mechanisms for enforcement and avenues for legal redress, on national and international levels. Thus citizens need to be educated in the safeguarding and practice of their human rights.
- Future social harmony likely depends on sensitizing young citizens before biases and prejudices are formed and hardened. (Urman, S. 1988).

The United States (signatory to the *Universal Declaration of Human Rights* but not to the two Covenants) claim a national tradition of concern about human rights. The *Declaration of Independence* asserts that the government of the United States was founded on the explicit belief that the primary purpose of government is to secure and protect the rights of its citizens, while the first ten amendments to the *Constitution* (the *Bill of Rights*) compensate for the lack of specificity of that document. The last quarter century has witnessed phenomenal movement in the direction of implementing civil rights legislation at both federal and state levels and at least eight states have, at this time, either produced curriculum and/or mandated the teaching of human rights. Rationale for human rights education in the U.S. has been provided by two prominent educators:
- To illuminate the human condition and stress the universality of the search for human dignity.
- To develop effective citizens who will have an impact on the policies of their government, and who have an understanding of the foreign policy of their own and other governments.
- To serve as an organizing framework for global studies which otherwise can lack substance and focus.
- To dissipate students' egocentric and ethnocentric view of rights — implying responsibilities toward others (Branson, M. and Torney-Purta, J. 1982).

Alternative conceptions of human rights education in the eastern bloc countries and in Third World nations must also be recognized. Comparing human rights education programmes in the socialist state of the German

---

[5] Note UNESCO Conventions and United Nations Covenants are international treaties, passed by a two-thirds majority of the General Conference, and which are legally binding for those states that ratify them. The *International Covenant of Economic, Social and Cultural Rights* (signed by eighty-three nations) calls for education directed to strengthening respect for human rights and fundamental freedoms. The *International Covenant on Civic and Political Rights* (signed by eighty nations) proclaims the rights of ethnic, religious or linguistic minorities 'to enjoy their own culture, to profess and practice their own religions, or to use their own language'.

Democratic Republic (GDR), the western state of the Federal Republic of Germany (FRG), and the developing nation of Zimbabwe, Shafer notes that in the GDR "... human rights are viewed as the collective rights of citizens of the state, while in the FRG as in other Western democracies, the state is charged with the preservation of individual rights, i.e., human rights. Human rights education in schools varies accordingly. In Third World countries, such as Zimbabwe, a fusion between the collectivist and the individualist approach to human rights education has not yet occurred (Shafer, 1987, p. 203).

Most developing countries, dedicated first and foremost, to programmes of national development, grant low priority to the teaching of human rights (Diokno, 1983). In fact, many governments hesitate to introduce this subject into their curricula for fear of the implications for social change (de Senarclens, 1983). Selle-Hosbach (1987) points out that political socialization operates within national systems with varying values and that a strong nationalist outlook is often accompanied by a mistrust of international organizations and their goals. For those nations which subscribe to international agreements dedicated to guaranteeing human rights and to the principles of democratic forms of government, the teaching of human rights is, however, both a moral and legal imperative (Urman, 1986).

2.5.4 WHERE AND WHEN DOES HUMAN RIGHTS EDUCATION FIT IN THE CURRICULUM?

Human rights are not an isolated 'extra', something to be stuffed into an already packed curriculum, but a concept that is integral to existing curriculum. Teaching about international human rights in our secondary schools is as necessary as teaching about who we are and the nature of the world we live in. (King, D., and King, S., 1982, p. 69).

Where? As implied above, teachers, the world over, are already complaining about not enough hours in the day to include all they are supposed to teach. Where can they fit in another 'subject'? In most primary and secondary schools, human rights is not treated as a separate subject, but rather infiltrates such courses and programmes as social studies, music, art, literature, language, philosophy, ethics, history, geography, economics, and political science, as appropriate. At its best, it finds 'windows of opportunity' to permeate the framework of the formal as well as the informal curriculum of primary and secondary schools. Extracurricular activities, the teaching-learning strategies of the classroom, and the entire ethos and climate of the school are part of the human rights "curriculum".

When? There are alternate conceptions of the optimum time for introducing human rights education. A rationale and recommendations for including human rights education at *preschool* level is provided in *UNESCO Recommendation* (1974) §24 and supported by the Fortieth Council of Europe Teachers' Seminar (Abdallah-Pretceille, M., 1989). Humphrey (1987) cautions that it is important to get the human rights message across to students when they are young and before they have absorbed prejudices. His advise is "Aim for *primary grades*". Torney-Purta notes that, on the basis of the *primary principle*, the earlier an experience takes place in a child's life, the more formative it is likely to be. She points out that proponents of this principle are countered by those who subscribe to the *recency principle* (e.g. postponing instruction in the field until pupils are able to see direct links to voting, election campaigns, etc.). Opting for the *plasticity principle*, Torney-Purta points out that middle childhood (ages seven to eleven) appears to be the best time for formally introducing human rights curriculum. At this stage, a variety of important cognitive competencies have been achieved but many concepts and attitudes are not yet rigid or fixed.

At about the age of seven, many children enter a period of rapid social-cognitive development and achieve at least rudimentary perspective-taking and concrete analytic skills. Although cognitive growth continues through adolescence, at abut the age of thirteen or fourteen there appears to be a lessening of attitudinal plasticity and increased stereotyping. Attitudes become rigid, and they are used more frequently as a way of confirming peer group solidarity and excluding those who are different. The equitable treatment of others, especially those who may be victims of injustice or lack of opportunity, seems to be less important than maintaining the superior position of the in-group during adolescence. (Torney-Purta, 1982, p. 43).

Noting that political views and consciousness become differentiated between the ages of eleven and thirteen more markedly than at any time before or after, Selle-Hosbach (1987) suggests that education involving political socialization must begin at *preschool or primary level.*

Thus, it appears that from early beginnings as a secondary school subject, human rights has made its way *down* the educational ladder to early adolescence, middle childhood, and even early childhood. A number of documents referred to above have stressed the *longitudinal approach* to human rights education starting with an informal approach and emphasis on process in preschool and primary levels, through more content-oriented programmes at the secondary level. University courses, adult education and, of course, teacher training are also seen as appropriate targets for human rights education.

2.5.5 HOW DOES HUMAN RIGHTS EDUCATION RELATE TO OTHER PROGRAMMES?

Historically, human rights education can serve as the common thread and logical outcome of the sequence of educational efforts responding to the increased pluralization of western democratic societies. This response, in the context of early twentieth century assimilationist goals, produced the *intergroup education* movement, stressing tolerance and mutual understanding. As assimilationist goals gave way to the valuing of cultural pluralism, several successive educational responses have dominated the latter half of the century. First, *ethnic studies* programmes, geared primarily to members of various ethnic and racial groups, focused on famous personalities, cultural traditions, and ethnic pride. (Later, *bilingual education* or *minority language* programmes added an emphasis on the rights of different ethnic groups to be educated in their primary language.) Recognising the need for all children to understand cultures other than their own, *multicultural education* programmes became the byword of the 70s. Early programmes took the "Tacos on Tuesday" approach but, more recently, there has been an emphasis on more general, universally appropriate concepts and *anti-racist* or *prejudice reduction programmes*.[6] Since each of these successive educational responses is premised on the (human) rights of minority groups, the content and process of human rights education appears to be the logical ultimate outcome of these movements (See Figure 1).

Shafer (1987) points out the relationship between human rights education and three other areas of curriculum emphasis - *global education, moral education,* and *social and civic education.* Tucker (1982) notes three major drives competing for attention (*global education, multicultural education,* and *civic education*) - all aiming to "improve our capacities for living humanely and justly with one another". He points out that one of the criticisms against global education is that it lacks a content base but that international human rights is one strand of content within global education that has developed a substantial mass of practice and scholarship, which, in

---

[6] The Council of Europe has drawn a distinction between the terms "multicultural" and "intercultural". "Multicultural" is used to describe a situation - of pluralist societies on regional, national and international levels. While "multicultural education" has often focused on teaching about heroes, holidays, foods, customs and traditions of other cultures, intercultural education moves beyond the teaching "about" other cultures to the level of interaction and interchange between cultures-to the mutual enrichment of each.

194

Figure 1

Evolution of Educational Responses to The Needs of Pluralist Societies

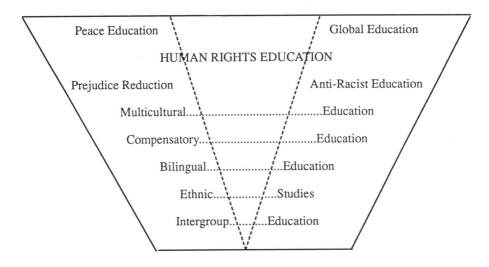

effect can add academic credibility to the latter. The concern of proponents of multicultural education, or *intercultural education*, (the preferred term of the Council of Europe) for the rights of minority cultures within a country, should not preclude attention to the rights of minority cultures on a world-wide basis. *Civic and law-related education* stress human rights, often *only* within the context of national history and government. But, as Branson and Torney-Purta point out, "if students study human rights in the national setting only, they will develop a narrow perspective, one which may not serve them very well as adult citizens in an increasingly interdependent world." (Branson and Torney-Purta, 1982, p. 2). Thus, stressing one's responsibilities as a citizen of the world need not threaten one's loyalty, patriotism, or responsibilities to one's country. Butts views human rights education as the connecting link between *civic education* and *global education*, each of which has attracted zealous supporters in the 80s.

I strongly urge social studies educators to increase the extent to which education for civic cohesion and education for cultural pluralism are linked together and interwoven with education for global interdependence. I believe that this can be done especially well through emphasis upon international human rights (Butts, 1982, p. 25). Internationalists and liberal groups press for more attention to global issues, bemoaning the inadequate preparation of students for a globally interdependent world. Conservative groups clamour for greater attention to patriotic values and education for citizenship in a democracy. Representatives of various national, ethnic, racial and religious groups espouse multicultural (or intercultural) education. *Environmental education* (concerned with the need to conserve and the right to share in the world's resources), *peace education* (concerned with the right to live together free from the scourges of war) and *development education* (concerned with Third World issues) as well as *prejudice reduction* or *anti-racist education programmes* are all vying for space in the curriculum. It appears eminently clear that there is no possibility of all of these being implemented as add-ons to the curriculum. Since the principles of human rights permeate each of these programmes, however, a focus on the process and content of *human rights education* would assure attention to the critical aspects of each. Thus, human rights education can serve as the unifying factor which cuts across current efforts to produce informed and active citizens of their communities, their nations and of an interdependent world.

Figure 2. Human Rights Education in Relation to Current Educational Responses to the Need for Informed and Active Citizens

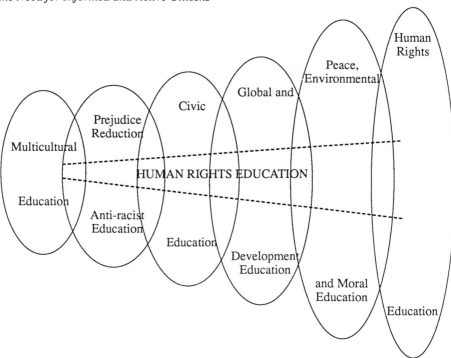

2.5.6 WHAT ARE SOME OF THE PROBLEMS IN IMPLEMENTING HUMAN RIGHTS EDUCATION IN THE SCHOOLS?

Although, as we have seen, the conceptualization of human rights programmes varies, there appear to be problems in achieving successful implementation common to all, or most, of these programmes. Starkey (1984) points out an educational paradox. Usually new curricula or innovations struggle to achieve recognition and legitimization. Human rights education, however, has both recognition and legitimization at the highest levels of government in a variety of international fora. Yet, for the most part, teachers have been unable or unwilling to find appropriate ways of introducing the subject in their classrooms on a regular basis. This section will examine some of the possible explanations for this phenomenon.

Perhaps one of the problems is the need for supporters of human rights education to make an impact at national and regional (rather than international) levels. It is a well-known phenomenon that *curriculum coming "from the top down"* (administrators to teachers) has less chance of being effectively implemented than if teachers, themselves, are actively involved in its development and committed to its implementation, - in effect, feeling a *sense of ownership of a project*. This same phenomenon may be operative on the part of national and regional officials in relation to projects initiated at international levels. The *level of importance attached by the society to human rights* is a contributing factor. "Where there is a genuine commitment or sense of urgency, an effort is made to include them (human rights) in the curriculum... where human rights are repeatedly curtailed ... human rights education has been absent from the curriculum". (Shafer, 1987, p. 203). In some cases, *hazy borders between federal/state or provincial/local responsibility* for policies and programmes result in sporadic implementation only on the basis of individual teacher initiative.

The *paucity of appropriate educational materials* is a reason given by many educators for their failure to deal with global issues and human rights (Buergenthal and Torney, 1976, p. 17). Drubay (1981) points out that the abstract legal language of documents has to be decoded in order to place them within linguistic grasp of students, and that materials need to be developed to help students deal with the contradictions, violations, and even hypocrisy that they sense in terms of world reaction to human rights issues. Materials for the middle grades are especially scarce - a fact recognised by the Canadian Human Rights Foundation, which, considering this age level the optimal period for beginning formal human rights education, is in the process of developing and testing such materials. They are also aware that the production and dissemination of attractive and worthwhile educational materials does not guarantee their use and are providing a significant component of in-service teacher training and ongoing support in the use of these materials.

The *in-service training of teachers* and *continuing, ongoing support for those working in the field* are absolutely necessary to ensure successful implementation and avoid having excellent materials languish on the classroom shelf. Teachers need to be helped to understand that the content area of human rights must be accompanied by a classroom environment and strategies based on human rights principles. Often teachers are concerned about *how to fit human rights education into the curriculum* and do not understand that human rights content is not a curriculum add-on, but a means of unifying and integrating existing frameworks. They need guidance and opportunities to work as teams to find these "windows of opportunity" in the existing curriculum frameworks. A network of those involved in development and implementation of human rights curriculum is lacking - could be an invaluable support system.[7]

The *competition of the various social educational movements* literally bombards teachers and school administrators with "critical" programmes and proposals as each lobbies for a place in the school curriculum. The overlapping nature of all of these programmes has not been acknowledged nor the need for their co-operation and unification met.

Attention must also be directed at the *inadequate pre-service preparation of teachers* in both the content and processes of human rights education.

Although peripheral in most national teacher education efforts, how to prepare teachers to understand and appreciate other cultures in a human rights context has been central in UNESCO programmes (yet) in both UNESCO and national reports, the most frequently noted shortcoming in efforts to advance human rights teaching, at all levels and in all fields, is the lack of well-qualified teachers (Sebaly, 1987, p. 209).

Teachers are generally unwilling to risk teaching on topics or in areas where they feel ill-prepared. Yet, the *academic preparation* of future teachers rarely includes such topics as the history of human rights, international instruments and enforcement mechanisms, important persons and movements in the field, and current human rights issues. Likewise,

---

[7] The author is familiar with an extraordinarily successful network of this type in the field of global education (in California). BAGEP (the Bay Area Global Education Program) has created a cadre of committed teachers, actively involved in curriculum development and implementation and who continue to draw others into the network.

their *pedagogic preparation* rarely provides practical experience with the recommended processes of active, cooperative and small group learning, critical thinking, problem solving, and conflict resolution, simulation and role playing - all alternative conceptions to the frontal-type teaching which still constitutes the major portion of the experiences of future teachers in teacher-training institutions.[8]

Another problem stems from *the controversial nature of the subject matter* itself (Urman, 1988). Where school administrators have the authority to include the subject of human rights in the curriculum of their schools, many are concerned about community reaction. There is resistance from conservative constituents who fail to see the link between civic education for national values and education for citizenship in an interdependent world. Often a committed teacher has to convince a skeptical administrator, who has to deal with a hostile school board. Introduction of this type of curriculum requires a *co-operative effort proceeding both from the top* (school boards and administrators) in order to get official approval, *as well as from the bottom* (teachers and pupils, in order to assure implementation once approval is obtained. And, *teachers need to be assured of protection* if they are to deal with controversial issues in the classroom, Drubay, 1981, Starkey, 1985, Shafer, 1987)

Better formative and summative evaluation tools and findings could make a stronger case for human rights education. Torney-Purta (1982) points out that the research base for human rights education is not as strong as it should be. She believes *further research* is needed to address such issues as what young people know about human rights and what misconceptions they hold, the best sequence and timing of instruction in human rights, plus evaluation of the effectiveness of informal learning experience (school climate, etc.), teaching materials, and methods.

2.5.7 CONCLUSIONS

An attempt has been made to offer a definition of human rights education appropriate to the array of value systems, forms of government and approaches to human rights reflecting the world system as we come to the close of its twentieth century. Although evolving from an acceptance of the Universal Declaration of Human Rights (and its principles) as an indisputable part of the customary law of nations, it recognizes that

---

[8] Interviews conducted and questionaires administered in two teacher training institutions in different regions of Spain, by this author, indicated that nowhere in the three year curriculum was any attention given either to the content or processes described above.

different societies define rights in terms of their own historical experience, their value systems, and the political and economic realities of the current era. Thus, the form and content of human rights education will depend on the values into which societies wish to socialize their children.

The components of human rights education, as practised in different societies, have been classified and described according to a schema that appears to be appropriate and acceptable to a wide variety of conceptualizations. Included are *content* (including knowledge, skills, values and social participation and *process* variables (including all aspects of the ethos and environment of the classroom and the school).

Major attention was focused on the 1974 UNESCO *Recommendation concerning Education for International Co-operation and Peace and Education relating to Human Rights and Fundamental Freedoms* as the rationale for the inclusion of human rights education in the curricula of all member states. For those nations belonging to the Council of Europe, additional emphasis was placed on *Resolution (78) 21 on the Teaching of Human Rights*, and on *Recommendation No. R (85) 7 on Teaching and Learning about Human Rights in Schools*, both adopted by the Committee of Ministers of the Council of Europe, as additional rationale for inclusion of human rights education in the curricula of all their member states. As examples of legislative rationale on a national level, Canadian and U.S. documents were reviewed, and finally, rationale on moral and educational bases were cited to establish the basis for human rights education in all societies.

Alternative conceptions of "where" and "when" human rights education should be part of the curriculum were discussed. The consensus appears to be that such programmes should not be "add-ons" to the curriculum. They can best begin informally as early as preschool, with formal topics and materials introduced in the upper elementary years, continuing through secondary school as a component of various traditional courses. In relation to other current proposed programmes in the area of social education, it was suggested that human rights education can serve as the unifying factor which cuts across current efforts to produce informed and active citizens.

Finally, it was noted that, in spite of recognition and support at some of the highest levels, human rights education programmes are not gaining widespread acceptance and implementation in the schools. Various possible explanations for this phenomenon were reviewed. It remains for succeeding colloquies to deal with the task of suggesting how best to create a core of committed, confident and competent teachers prepared to implement human rights education within a classroom and school environment in which human rights are ever present and respected.

2.5.8 BIBLIOGRAPHY

ABDALLAH-PRETCEILLE, M. (1989) Human Rights Education in Pre-primary Schools. Report of the 40th Council of Europe Teachers' Seminar Donaueschingen, 20-25 June, 1988, Strasbourg, Council of Europe.

BRANSON, M. and TORNEY-PURTA, J. Introduction, in BRANSON, M. and TORNEY-PURTA, J. (Eds.) International Human Rights. Society and the Schools. Washington, National Council for the Social Studies, 1982, pp. 1-7.

BUERGENTHAL, T. and TORNEY, J. International Human Rights and International Education, Washington, U.S. National Commission for UNESCO, 1976.

BUTTS, R.F. International human rights and civic education, in: BRANSON, M. and TORNEY-PURTA, J. (Eds.) International Human Rights, Society and the Schools. Washington, National Council for the Social Studies, 1982, pp. 23-34.

Committee of Experts for the Promotion of Education and Information in the field of Human Rights (DH-ED) (17 June 1987), Education and Information in the Field of Human Rights: Activities of the Council of Europe. Strasbourg, Council of Europe.

COOMARASWAMY, R. A third world view of human rights. UNESCO Courier, August/September, 1982, pp. 49-50.

CROMBIE, Human rights education: Canadian and international responsibilities, in: Human Rights Education in Canada, Montreal, Canadian Human Rights Foundation, 1988.

DE SENARCLENS, P. Research and teaching of human rights: introductory remarks, in EIDE, A. and THEE, M. (Eds.) Frontiers of human rights education. Oslo, Universitetsforlaget, 1983.

DIOKNO, J. Human rights teaching and research in the context of development: the east-west and north-south dimensions, in: EIDE, A. and THEE, M. (Eds.) Frontiers of human rights education, Oslo, Universitetsforlaget, 1983.

DRUBAY, A. Human rights instruction and education in secondary schools, Strasbourg, Council of Europe, 1981.

EIDE, A. Pocket guide to the Development of Human Rights Institutions and Mechanisms, Strasbourg, Council of Europe, 1989.

HUMPHREY, J. Epilogue, in: TARROW, N. (Ed.) Human rigths and education, Oxford, Pergamon, 1987, pp. 235-236.

HUMPHREY, J. Human rights and the United Nations: a great adventure. Dobbs Ferry, NJ. Transnational, 1984.

KING, D. and KING, S. Teaching human rights in secondary schools, in: BRANSON, M. and TORNEY-PURTA, J. (Eds.) International Human

Rights, Society and the Schools. Washington, National Council for the Social Studies, 1982, pp. 61-70.

Learning for Life Strasbourg, Directorate of Press and Information of the Council of Europe, 1984.

NCSS (National Committee for the Social Studies) Curriculum Guidelines for the Social Studies. Washington, NCSS, 1979.

OBERMEIER, K. Human rights: an international linguistic hyperbole, in: SCHWEDA-NICHOLSON, N. (Ed.) Languages in the International Perspective. Norwood, NJ. Ablex, 1986, pp. 105-114.

RAY, D. and TARROW, N. Human rights and education: an overview, in: TARROW, N. (Ed.) Human rights and education. Oxford, Pergamon, 1987, pp. 3-16.

SEBALY, K. Education about human rights: teacher preparation, in: TARROW, N. Human rights and education. Oxford, Pergamon, 1987, pp. 207-222.

SELLE-HOSBACH, K. Teachers Seminar on Human Rights Education and the Teaching of Social, Civic and Political Education. Donaueschingen, 17-21 November, 1986, Strasbourg, Council of Europe, 28 September 1987.

SHAFER, S. Human rights education in schools, in: TARROW, N. (Ed.) Human rights and education. Oxford, Pergamon, 1987, pp. 191-206.

STARKEY, H. Teachers' course on teaching and learning about human rights in schools. Strasbourg, Council of Europe, 1985.

STARKEY, H. Symposium on Human Rights Education in Schools in Western Europe. Vienna, 17-20 May 1983. Strasbourg, Council of Europe, 1984.

STOBART, M. Prologue, in: TARROW, N. (Ed.) Human rights and education. Oxford, Pergamon, 1987.

TARROW, N. Human rights education: a comparison of Canadian and U.S. approaches, paper presented at the Comparative and International Education Society, Western Region Conference, Sacramento, October 1988.

TORNEY-PURTA, J. Human Rights Education: A Rationale and Research Evidence, in: Human Rights Education in Canada, Montreal, Canadian Human Rights Foundation, 1988.

TORNEY-PURTA, J. Socialization and human rights research: implications for teachers, in: BRANSON, M. and TORNEY-PURTA, J. (Eds.) International Human Rights, Society and the Schools. Washington, National Council for the Social Studies, 1982. pp. 35-48.

TUCKER, J. International human rights in secondary schools, in: BRANSON, M. and TORNEY-PURTA, J. (Eds.) International Human Rights, Society and the Schools. Washington, National Council for the Social Studies, 1982, pp. 71-80.

URMAN, S. A National initiative to increase the teaching of human rights in Canadian schools, in: Human Rights Education in Canada, Montreal, Canadian Human Rights Foundation, 1988.

URMAN, S. Human rights education: a legal and moral imperative, Canadian Journal of Education 11, 1986, No. 3, pp. 383-387.

Bibliography compiled by the Council of Europe

UNESCO

Rapport concernant les activités destinées à donner effet aux recommendations de la Conférence intergouvernementale sur l'éducation pour la compréhension, la coopération et la paix internationales et l'éducation relative aux droits de l'homme et aux libertés fondamentales, en vue de développer un état d'esprit favorable au renforcement de la sécurité et au désarmement (1983), UNESCO, Paris, Septembre 1987, 23 pages, référence 24 C/92.

L'UNESCO et l'éducation pour la paix (UNESCO and peace education) in: Perspectives, Vol. XV, No. 3, 1985, pp. 487-496. Paris, UNESCO, 1985, 10 pages.

Robert ASPESLAGH & Adrian NASTASE, Educational and Research Programmes of Universities and Research Institues in the European Region devoted to International Understanding, Co-operation, Peace and Respect for *Human Rights*, UNESCO Centre for Higher Education (CEPES), Stirbey Voda, Bucarest 1987, 88 pages.

FIPESO
(Fédération Internationale des Professeurs d'Enseignement Secondaire Officiel)
Congrès de la FIPESO, Lisbonne, 24-28 juillet 1984. Dans une société de transformation avec un haut niveau de chomage, comment le secondaire peut-il *préparer les jeunes* aux exigences de la *vie sociale* et culturelle, in: L'Athenée, No. 5, Nov./Dec. 1984, pp. 61-66, Liège: Femo, 1984, 6 pages.
(A summary of the FIPESO Congress is also given in Bulletin International, Genève, FIPESO, Jan. 1985, 10 pages.)

AUSTRIA

Peter LEUFRECHT, Der Europarat und die Menschenrechte (The Council of Europe and the Human Rights), Wien, Verlag für Geschichte und politik, 1987, 28 pages.
ÖSTERR. INSTITUT FÜR FRIEDENSFORSCHUNG UND FRIEDENS-ERZIEHUNG, Friedens-Forum, Hefte zur Friedenserziehung (Peace-Forum, Journal on Peace Education), Burg Schlaining, No. 3/1988, pp. 11 seq., Stadtschlaining, Burgenland.
Günther OGRIS, Das Weltbild der 14-järighen und ihre Sichtweise des Nord-Süd-Konflikte (What 14 year-old adolescents think of the world and the North-South conflict), Wien, Institut für empirische Sozialforschung, 1988, 66 pages.
Günther OGRIS, Einflüsse auf den Entwicklungspolitik-Unterricht (Factors influencing education for development policy), Wien, Institut für empirische Sozialforschung, 1988, 53 pages.

BELGIUM

William DE COSTER et al. L'echelle des valeurs des adolescents flamands (The value scale of Flemish adolescents), in: Scientia Paedagogica Experimentalis, Vol. XXV, No. 2, 1988.
William DE COSTER, Eric POT & Geert DE SUETE, Value Hierarchies in Flemish Adolescents and Parents, in: Psychologica Belgica, Vol. 27, 1987, No. 2.

CANADA

CENTRALE D'ENSEIGNEMENT DU QUEBEC, Déclarer la paix,

réclamer nos droits (Declaring peace, claiming our rights), September 1988, Doc. D-9183.

MINISTERE DE LÉDUCATION (1035 Rue De La Chevrotière, Quebec GIR 5A5), Guide pédagogique et de référence, La tournée internationale "jeunesse pour la paix et la justice" (Educational and reference guide, International Tour "Youth for Peace and Justice), 1988, 133 pages.

CHINA

ALL-CHINA YOUTH FEDERATION, What does love, marriage and sex mean to Chinese youth? Beijing, Chinese Youth, Vol. XII, No. 2. 1989, pp. 38.

CZECHOSLOVAKIA

Jarmila KOTÁSKOVÁ', Socializace a moralni vývoj díťéte (Socialisation and moral development in children) with summary in English, Praha, Academia nakladatelství' Ceskoslovenske akademieved 1987. Studie CSAV, 9-87, 200 pages.

DENMARK

THE LABORATORY OF DEMOCRATIC EDUCATIONAL RESEARCH, (Danmarks Laererhøjskole, Emdrupvej 101, DK-2400 KÖBENHAVN NV), Newsletter 2, European Network toward school democratisation, Vol. 1, October 1988, No. 2.

Sven Erik NORDENBO, Children's Rights, die Antipädagogen and the laternalism of John Stuart Mill, in: Scandinavian Journal of Educational Research, Vol. 31, 1987, 163-180.

Sven Erik NORDENBO, Justification of laternalism in education, in: Scandinavian Journal of Educational Research, Vol. 30, 1986, pp. 121-139.

FRANCE

CNDP, Pour une éducation aux droits de l'homme: ressources documentaires collectées et commentées par une équipe d'enseignants et de chercheurs de l'Institut National de Recherche Pédagogique. Paris:

Centre National de Documentation Pédagogique, 1985, 88 p.: III. (Références documentaires; 30).
Guide pratique indiquant aux enseignements les principales références écrites et audiovisuelles, utilisables par les enseignants pour leur formation et leur enseignement dans le domaine des droits de l'homme. Ils'agit de documents en langue française, d'un accès aise, qui abordent la notion de droits de l'homme par tous ses aspects (fondements et aspects spécifiques). En annexe figurent des textes officiels et des adresses utiles.

Jacques COLOMB and François AUDIGIER, Pour une éducation aux droits de l'homme (Human Rights Education), report on the results of an INRP research and development project, C.N.D.P., Paris, novembre 1985, 88 pages, ISSN 0761-5264.

FEDERAL REPUBLIC OF GERMANY

FREIE HANSESTADT BREMEN, WISSENSCHAFTLICHES INSTITUT FÜR SCHULPRAXIS (Am Weidedamme 20, D-2800 BREMEN), Arbeitsberichte Folge 8, 1978, Lehrerqualifikationen für die Sekundarstufe 1 (Teacher qualifications for lower secondary education, including value education).

Joachim DETJEN, Grundpflichten als Erziehungsaufgabe der Schule, in: Pädagogische Welt, Vol. 42, 1988, No. 3, pp 14-118.

DEUTSCHER BUNDESTAG, Förderung der Menschenrechtserziehung, Gemeinsame Entschliessung der Fraktionen des Deutschen Bundestages (Promotion of Human Rights Education, Joint Resolution of all Groups of the German Federal Parliament), in: UNESCO-Dienst, Vol. 27, 1980, No. 5, pp 7-8.

Peter DOBBELSTEIN-OSTHOFF & Heinz SCHIRP, Werteerziehung in der Schule - aber wie? Ansätze zur Entwicklung der moralisch-demokratischen Urteilsfähigkeit (Value education at school - but how? Attempts to develop moral and democratic judgment). Landesinstitut für Schule und Weiterbildung, Curriculumentwicklung in Nordrhein-Westfalen, Soest (Paradieser Weg 64, D-4770 SOEST, 1987, 109 pages.

Jürgen HAMBRINK, Menschenrechtsentziehung - deutsche Beiträge zu einem internationalen Programm (Human rights education - German contributions to an international programme), in: Bildung konkret, Vol. 12, 1981, No. 4, pp. 8-9.

Wolfgang HEINZ, Menschenrechtserziehung tut not (Human rights education is necessary), in: Liberal, Vol. 23, 198, No. 12, pp. 88-883

Wolfgang HÜFNER & Hans-Wolf RISSOM, UNESCO, die Bundesrepublik Deutschland und die internationale Erziehung (UNESCO, the Federal Republic of Germany and International Education), in: Zeitschrift für internationale erziehungs- und sozialwissenschaftliche Forschung, Vol. 3, 1986, No. 2, pp. 193-216.

Gottfried KLEINSCHMIDT, Dokumente, Publikationen Berichte, Bücher zur Werterziehung in der Schule und zu angrenzenden Bereichen, Zusammenstellung und Kommentierung (Documents, publications, reports and books about the value education at school and related areas compiled and documented), Landesinstitut für Erziehung und Unterricht, Rotebühlstrasse 133, D-7000 Stuttgart 1, 249 pages, 1988.

Dieter KOCH, Christen in politischen Konflikten des 20. Jahrhunderts, Didaktische überlegungen und Materalen zu einem Unterrichtsvorhaben (Christians in political conflicts throughout the 20th century, didactical approaches and material for a teaching project), Arbeitsberichte (working papers) No. 39 and 40, 1985, published by Wissenschaftliches Institut für Schulpraxis, Am Weidedamme 20, D-2800 Bremen 1, 59 pages and 35 pages.

KULTUSMINISTERKONFERENZ (Postfach 2240, D-5300 Bonn 1), Empfehlung zur Förderung der Menschenrechtserziehung in der Schule (Recommendation of the Conference of German Ministers of Education on the Promotion of Human Rights Education at School), Bonn, 4 Dec, 1980, 3 pages.

KULTUSMINISTERKONFERENZ (Postfach 2240, D-5300 Bonn 1), Empfehlung zur Förderung der Menschenrechtserziehung in der Schule (Recommendation of the Conference of German Ministers of Education on the Promotion of Human Rights Education at School), Bonn, 4 Dec. 1980, 3 pages.

MENSCHENRECHTE in einer sich wandelnden Welt. Ein Lehrerpreisausschreiben in der Bundesrepublik Deutschland (Human Rights in a changing world. A competitive award for teachers in the Federal Republic of Germany), in: UNESCO-Dienst, Vol. 26, 1979, No. 3, pp. 7-10.

SCHLESWIG-HOLSTEINISCHER LANDTAG (Parliament of Schleswig-Holstein), Drucksache 10/1627 of 7 July 1986 on South Africa and human rights education at school (summing up present government policy based on value and human rights education in all subjects).

STAATSINSTITUT FÜR SCHULPÄDAGOGIK UND BILDUNGS-FORSCHUNG (Arabellastrasse 1, D-8000 München 81)

1. Oberste Bildungsziele in Bayern, Artikel 131 der Bayerischen Verfassung in aktueller pädagogischer Sicht (Supreme educational aims in Bavaria, Current educational approach to Article 131 of the Bavarian Constitution, Munich, 3 Sept. 1988, 23 pages.

2. Was ist Frieden? Handreichung zur Behandlung des Themas "Frieden" im Unterricht; Teil I: Grundlagen; Teil II: Unterrichtspraktische Hinweise (What is peace? Guide for treating the theme of peace at school. Part I: Basic problems; Part II: Practical advice for teachers), Munich, Oct. 1983 and Jan. 1984, Part I: 222 pages; Part II: 206 pages.

LUXEMBOURG

Ministère de l'Education Nationale et de la Jeunesse, 6 Boulevard Royal, L-2910 Luxembourg.
Programmes de formation morale et sociale (Curriculum for moral and social education: values, human rights, democracy, socialisation).

THE NETHERLANDS

See also separate bibliography prepared by SVO (Institue for Educational Research in the Netherlands, Sweelinckplein 14, NL-2517 GK The Hague).
Hans HOOGHOF, Curriculum Development for Political Education in the Netherlands, National Institute for Curriculum Development (P.O. Box 2041, NL-7500 CA Enschede), May 1987, 22 pages.

SWITZERLAND

Jacques-A. TSCHOUMY, Les droits de l'homme, une éducation civique et morale pour notre temps? Elaboré à l'occasion du 40e anniversaire de la Declaration universelle des Droits de l'Homme (Human Rights - Civic and moral education for our present time? Prepared on the occasion of the 40th anniversary of UNESCO's Universal Declaration of Human Rights). Institut Romand de Recherche et de Documentation Pédagogiques, Neuchâtel, 1988, 29 pages.

UNITED KINGDOM

Roger AUSTIN, From conflict to cooperation, a European studies (Ireland and Great Britain) project, Irish Institute for European Affairs; (Folk and Transport Museum, 153 Bangor Road, Cultra, Co. Down, BT 18 OEU), Project Handbook 988-989.
CENTRE FOR THE STUDY OF CONFLICT, Annual Report to Research

Committee, Senate and Council, University of Ulster, Cromore Road, Coleraine, Co. Londonderry BT 52 1SA.

J. MACBEATH, D. MEARNS, W. THOMSON & S. HAW, Social Education, The Scottish Approach (Phase 1 report), Jordanhill College of Education, Glasgow, 1981.

J. MACBEATH, D. MEARNS, H. RODGER & W. THOMSON, Social Education, A West German View, Jordanhill College of Education, Glasgow, 1981.

J. MACBEATH, D. MEARNS, H. RODGER & W. THOMSON, Social Education, Notes towards a Definition, Jordanhill College of Education, Glasgow, 1981.

J. MACBEATH, D. MEARNS, W. THOMSON & S. HAW, Social Education, Through Special Programmes, Jordanhill College of Education, Glasgow, 1982.

J. MACBEATH, J. GOLDIN & A. ROBERTSON, Social Education, Using Media, Jordanhill College of Education, Glasgow, 1983.

D. MEARNS, Inveralmond Alternative School, Evaluation Phase 1 (about social education), Jordanhill College of Education, Glasgow, 1982.

D. ORR & J. MACBEATH, Social Education: Through Outdoor Activities, Jordanhill College of Education, Glasgow, 1983.

YUGOSLAVIA

Dr. Edita SOS, Educational democratisation. Skolsk novine, Zagreb, May 1987, Faculty of Philosophy.

List of Participants

I. CHAIRMAN, RAPPORTEUR AND LECTURERS

Professor Dr. Manual FERREIRA PATRICIO (Président/Chairman).
Director, Ministerio da Educação e Cultura, Instituto de Inovaçao
Educacional, Travessa das Terras de Sant'Ana 15, P-1200 LISBOA

Mr. Hugh STARKEY (Rapporteur Général). Senior Lecturer, Westminster
College, GB - OXFORD OX2 9AT

M. François AUDIGIER. Institut National de Recherche Pédagogique,
29 rue d'Ulm, 75230 PARIS Cedex 05

Dr. Heinz SCHIRP. Landesinstitut für Schule und Weiterbildung,
Paradieser Weg 64, D-4770 SOEST

Dr. Bartolo PAIVA CAMPOS. Faculdade de Psicologia e de Ciencias de
Educação, Universidade do Porto, Rua das Taipas 76, P-4000 PORTO

Dr. Bengt THELIN. Swedish National Board of Education, Karlavägen
108, S-106 42 STOCKHOLM

M. Jacques-A. TSCHOUMY. Institut Romand de Recherche et de
Documentation Pédagogiques (IRDP). Faubourg de l'Hôpital 43,
CH-2007 NEUCHATEL 7

II. DELEGATES

BELGIUM
- Mr. Johan VANDERHOEVEN. Centre for Educational Policy and Innovation, Katholieke Universiteit, Vesaliusstraat 2, B-3000 LEUVEN

CYPRUS
- Mr. Doros THEODOULOU. Inspector, Secondary Education, 8 Klissouras, CY-NICOSIA 171

DENMARK
- Mr. Borge PRIEN. The Danish Institute for Educational Research, 28 Hermodsgade, DK-2200 COPENHAGEN

FINLAND
- Dr. Annikki JÄRVINEN. Institute for Educational Research, Seminaarikatu 15, SF-40100 JYVÄSKYLÄ

FRANCE
- Mme DUGAST. Directeur, Institut National de Recherche Pédagogique, 29 rue d'Ulm, PARIS Cedex 05

FEDERAL REPUBLIC OF GERMANY
- Dr. Erich HAPP. Direktor des Staatsinstituts für Schulpädagogik und Bildungsforschung, Arabellastrasse 1, D-8000 MÜNCHEN 81

GREECE
- Mme Myrto DRAGONA-MONAHOU, Professeur à l'Université de Crète, Présidente de l'Institut des Bourses d'Etat, 18 P Tsaldari Maroussi, GR-15122 ATHENS

HOLY SEE
- Rev. Père Guglielmo MALIZIA Professeur à la Faculté des Sciences de l'Education, Universita Pontificia Salesian, Piazza dell' Ateneo Salesiano 1, I-00139 ROMA

ICELAND
- Dr. Sigridur VALGEIRSDOTTIR. Director of the Icelandic Institute of Educational Research, Kennaraskolinn Fiusid, 101 Laufasvegi, ISL-101 REYKJAVIK

IRELAND
- Dr. Vincent GREANEY. Educational Research Centre, St. Patrick's College, Drumcondra, IRL-DUBLIN 9
Excused/excusé

ITALY
- Professor Antonio PAPISCA. Centre d'études et de formation sur les droits de l'homme et des peuples, Universita di Padova, Via del Santo 28, I-3513 PADOVA
- Dr. Marco MASCIA. Chercheur au Centre des Droits de l'Homme, Université de Padoue, Via del Santo 28, I-3513 PADOVA

LUXEMBOURG
- M. George WIRTEN. Directeur de l'Institut Supérieur d'Etudes et de Recherches pédagogiques, BP 2, L-7201 WALFERDANGE

MALTA
- Professor Kenneth WAIN. Senior Lecturer, Faculty of Education, University of Malta, Tal-Qroqq, M - MSIDA

NORWAY
- Mr. Per ØTERUD. Rector, Sagene Teacher Training College, Biermans gate 2, N-0473 OSLO 4

PORTUGAL
- Dr. Bartolo PAIVA CAMPOS. Faculdade de Psicologia e de Ciencias de Educaçao, Universidade do Porto, Rua das Taipas 76, P-4000 PORTO

SAN MARINO
- Mme Carla NICOLINI. Preside, Scuola Secondaria Superiore Statale, Contrada Santa Croce, RSM - 4703 CITTA DI SAN MARINO
- Mme Gemma CAVALLERI. Responsabile Ufficio Progetti Speciali, Dipartimento Istruzione e Cultura, Palazzo Begni, Contrada Omerelli, RSM - 47031 REPUBBLICA DI SAN MARINO

SWEDEN
- Dr. Bengt THELIN. Director of Education, Swedish National Board of Education, Karlavägen 108, S-106 42 STOCKHOLM
- Dr. Ingar MARKLUND. Director of Education, Swedish National Board of Education, S-106 42 STOCKHOLM

Excused/excusé

SWITZERLAND
- M. Jacques-A. TSCHOUMY. Directeur de l'Institut Romand de Recherche et de Documentation Pédagogiques (IRDP), Faubourg de l'Hôpital 43, CH-2007 NEUCHATEL

TURKEY
- Professeur Esin KAHIA. Faculté des Lettres, Histoire et Géographie, Aü Dil, Tarih, Cograpfya Fakultesi, Université d'Ankara, TR - ANKARA

UNITED KINGDOM
- Dr. Clare BURSTALL. Director, National Foundation for Educational Research in England and Wales, The Mere, Upton Park, GB - SLOUGH, Berks SL1 2DQ

YUGOSLAVIA
- Dr. Zlata GODLER. Professor of General Pedagogy, Faculty of Philosophy, University of Zagreb, Hrelinska 21, YU-41000 ZAGREB

III. OBSERVERS

COMMISSION OF THE EUROPEAN COMMUNITIES   Excused

OECD   Excused

WORLD CONFEDERATION OF ORGANISATIONS OF THE TEACHING PROFESSION (WCOTP)
- Mme Rita PERRAUDIN. SSPES, Docteur en philosophie et en lettres, 19 rue du Petit-chasseur, CH-1960 SION
- Mme Lurdes FERNANDES. FENPROF Av. Miguel Bombarda 61-8°, P-1000 LISBOA
- M. Antonio TEODORO. Secrétaire Général, FENPROF, Av. Miguel Bombara 61 - 8°, P-1000 LISBOA

INTERNATIONAL FEDERATION OF SECONDARY TEACHERS (FIPESO)
- M. Louis WEBER. 7 rue de Villersexel, 75007 PARIS

WORLD FEDERATION OF TEACHERS' UNIONS (FISE)
- M. Mario David SOARES. Trav. Infante Santo, 28, P-4435 RIOTINTO

OTHER OBSERVERS

PD Dr. Karl-Peter FRITZSCHE. Georg-Eckert Institut für Internationale Schulbuchforschung, Celler-Strasse 3, D-3300 BRAUNSCHWEIG

Mme Maria José CUEVA. Membre de la Junta de l'ERAIM, Equip de Recerca i d'Actuacio interculturals i sobre el Multilinguisme, Carrer Ferran Puig, 67, 2°, E-08023 BARCELONA

Mme Fernada OLIVEIRA. Agent national de liaison auprès de la Délégation du Conseil de la Coopération Culturelle (CDCC) du Conseil de l'Europe, Direction Générale de l'Enseignement primaire et secondaire (DGEBS), Av. 24 de Julho 140, P-1200 LISBOA

M. Joao Manuel COSTA E SILVA. Director de Servicos do Ensino Superior, Secretaria Regional da Educação, Juventude e Emprego, P-9000 FUNCHAL, MADEIRA

Prof. Dr. Norma TARROW. Professor of Education, Chair, Human Development Department, California State University, Long Beach, USA - CALIFORNIA 90840.

Mme Ana Maria de VERA CRUZ. Directora Adjunta do Gabinete de Estudos do Ministerio da Educação, REPUBLICA DE SAO TOME E PRINCIPE

M. Rui CORREIA LANDIM. Sociologue, Département des politiques d'éducation, Institut National pour le développement de l'éducation, Ministère de l'Education, de la Culture et des Sports, BP 132 BISSAU, REPUBLICA DE GUINEE-BISSAU

ORGANISERS

INSTITUTO DE INOVAÇÃO EDUCACIONAL

Professor Dr. Manuel FERREIRA PATRICIO. Director, Ministerio da Educação e Cultura, Instituto de Inovação Educacional, Travessa das Terras de Sant'Ana 15, P-1200 LISBOA

Dr. Teresa QUINTELA. Instituto de Inovação Educacional, Travessa das Terras de Sant'Ana 15, P-1200 LISBOA

Dra Graça Maria FERREIRA POMBA. Instituto da Educação Educacional, Travessa das Terras de Sant'Ana 15, P-1200 LISBOA

Dra Maria dos Anjos COHEN CASEIRO. Instituto da Inovação Educacional, Travessa das Terras de Sant'Ana 15, P-1200 LISBOA

Maria Emilia APOLINARIO. Instituto de Inovação Educacional, Travessa das Terras de Sant'Ana 15, P-1200 LISBOA

Barbara PALLA E CARMO. Av. do Uruguai, 36-1B, P-1500 LISBOA

COUNCIL OF EUROPE

Dr. Michael VORBECK. Chef de la Section de la Documentation et de la Recherche pédagogiques, Direction de l' Enseignement, de la Culture et du Sport, Conseil de l'Europe, BP 431 R6, F-67006 STRASBOURG Cedex
Mme Sylviane WEYL. Mme Danièle IMBERT. Section de la Documentation et de la Recherche pédagogiques, Direction de l'Enseignement, de la Culture et du Sport, Conseil de l'Europe,
BP 431 R6, F-67006 STRASBOURG Cedex

INTERPRETERS

Mme Christina FUTSCHER PEREIRA. Rue dos Caetanos, 5-4 esq., P-1200 LISBOA
Mme Sheilah CARDNO. Rue Presidente Arriaga, 34-1°, P-1200 LISBOA
Mlle Julia TANNER. Rua Angola 7, Resdo Chao Direito, P-1200 LISBOA
Mme Diana WYLDE. Via Palazzuolo 104, I-50123 FIRENZE

COUNCIL OF EUROPE

Recommendation on Human Rights Education

RECOMMENDATION No. R (85) 7

OF THE COMMITTEE OF MINISTERS TO MEMBER STATES ON TEACHING AND LEARNING ABOUT HUMAN RIGHTS IN SCHOOLS
(Adopted by the Committee of Ministers on 14 May 1985 at the 385th meeting of the Ministers' Deputies)

The Committee of Ministers, under the terms of Article 15.*b* of the Statute of the Council of Europe,

Considering that the aim of the Council of Europe is to achieve a greater unity between its members for the purpose of safeguarding and realising the ideals and principles which are their common heritage;

Reaffirming the human rights undertakings embodied in the United Nations' Universal Declaration of Human Rights, the Convention for the Protection of Human Rights and Fundamental Freedoms and the European Social Charter;

Having regard to the commitments to human rights education made by member states at international and European conferences in the last decade;

Recalling:
– its own Resolution (78) 41 on "The teaching of human rights",
– its Declaration on "Intolerance: a threat to democracy" of 14 May 1981,
– its Recommendation No. R (83) 13 on "The role of the secondary school in preparing young people for life";

Noting Recommendation 963 (1983) of the Consultative Assembly of the Council of Europe on "Cultural and educational means of reducing violence";

Conscious of the need to reaffirm democratic values in the face of:
– intolerance, acts of violence and terrorism;
– the re-emergence of the public expression of racist and xenophobic attitudes;
– the disillusionment of many young people in Europe, who are affected by the economic recession and aware of the continuing poverty and inequality in the world;

Believing, therefore, that, throughout their school career, all young people should learn about human rights as part of their preparation for life in a pluralistic democracy;

Convinced that schools are communities which can, and should, be an example of respect for the dignity of the individual and for difference, for tolerance, and for equality of opportunity,

I. Recommends that the governments of member states, having regard to their national education systems and to the legislative basis for them:
   a. encourage teaching and learning about human rights in schools in line with the suggestions contained in the appendix hereto;
   b. draw the attention of persons and bodies concerned with school education to the text of this recommendation;

II. Instructs the Secretary General to transmit this recommendation to the governments of those states party to the European Culture Convention which are not members of the Council of Europe.

APPENDIX TO RECOMMENDATION NO. R(85) 7

Suggestions for teaching and learning about human rights in schools

1. *Human rights in the school curriculum*

1.1. The understanding and experience of human rigths is an important element of the preparation of all young people for life in a democratic and pluralistic society. It is part of social and political education, and it involves intercultural and international understanding.

1.2. Concepts associated with human rights can, and should, be acquired from an early stage. For example, the non-violent resolution of conflict and respect for other people can already be experienced within the life of a pre-school or primary class.

1.3. Opportunities to introduce young people to more abstract notions of human rights, such as those involving and understanding of philosophical, political and legal concepts, will occur in the secondary school, in particular in such subjects as history, geography, social studies, moral and religious education, language and literature, current affairs and economics.

1.4. Human rights inevitably involve the domain of politics. Teaching about human rights should, therefore, always have international agreements and covenants as a point of referene, and teachers should take care to avoid imposing their personal convictions on their pupils and involving them in ideological struggles.

2. *Skills*

The skills associated with understanding and supporting human rights include:

i. *intellectual skills*, in particular:
- skills associated with written and oral expression, including the ability to listen and discuss, and to defend one's opinions;
- skills involving judgment, such as:
  - the collection and examination of material from various sources, including the mass media, and the ability to analyse it and to arrive at fair and balanced conclusions;
  - the identification of bias, prejudice, stereotypes and discrimination;

ii. *social skills*, in particular:
- recognising and accepting differences;
- establishing positive and non-oppressive personal relationships;
- resolving conflict in a non-violent way;
- taking responsibility;

- participating in decisions;
- understanding the use of the mechanisms for the protection of human rights at local, regional, European and world levels.

3. *Knowledge to be acquired in the study of human rights*

3.1. The study of human rights in schools will be approached in different ways according to the age and circumstances of the pupil and the particular situations of schools and education systems. Topics to be covered in learning about human rights could include:
    i. the main categories of human rights, duties, obligations and responsibilities;
    ii. the various forms of injustice, inequality and discrimination, including sexism and racism;
    iii. people, movements and key events, both successes and failures, in the historical and continuing struggle for human rights;
    iv. the main international declarations and conventions on human rights, such as the Universal Declaration of Human Rights and the Convention for the Protection of Human Rights and Fundamental Freedoms.
3.2. The emphasis in teaching and learning about human rights should be positive. Pupils may be led to feelings of powerlessness and discouragement when confronted with many examples of violation and negations of human rights. Instances of progress and success should be used.
3.3. The study of human rights in schools should lead to an understainding of, and sympathy for, the concepts of justice, equality, freedom, peace, dignity, rights and democracy. Such understanding should be both cognitive and based on experience and feelings. Schools should, thus, provide opportunities for pupils to experience affective involvement in human rights and to express their feelings through drama, art, music, creative writing and audiovisual media.

4. *The climate of the school*

4.1. Democracy is best learned in a democratic setting where participation is encouraged, where views can be expressed openly and discussed, where there is freedom of expression for pupils and teachers, and where there is fairness and justice. An appropriate climate is, therefore, an essential complement to effective learning about human rights.
4.2. Schools should encourage participation in their activities by parents and other members of the community. It may well be appropriate for

schools to work with non-governmental organisations which can provide information, case-studies and first-hand experience of successful campaigns for human rights and dignity.

4.3. Schools and teachers should attempt to be positive towards all their pupils, and recognise that all of their achievements are important - whether they be academic, artistic, musical, sporting or practical.

5. *Teacher training*

5.1. The initial training of teachers should prepare them for their future contribution to teaching about human rights in their schools. For example, future teachers should:
  i.   be encouraged to take an interest in national and world affairs;
  ii. have the chance of studying or working in a foreign country or a different environment;
  iii. be taught to identify and combat all forms of discrimination in schools and society and be encouraged to confront and overcome their own prejudices.

5.2. Future and practising teachers should be encouraged to familiarise themselves with:
  i.   the main international declarations and conventions on human rights;
  ii. the working and achievements of the international organisations which deal with the protection and promotion of human rights, for example through visits and study tours.

5.3. All teachers need, and should be given the opportunity, to update their knowledge and to learn new methods through in-service training. This could include the study of good practice in teaching about human rights, as well as the development of appropriate methods and materials.

6. *International Human Rights Day*

Schools and teacher training establishments should be encouraged to observe International Human Rights Day (10 December).